Mastering Surface Modeling with SOLIDWORKS 2023

Lani Tran

SDC
PUBLICATIONS

SDC Publications
P.O. Box 1334
Mission, KS 66222
913-262-2664
www.SDCpublications.com
Publisher: Stephen Schroff

ISBN-13: 978-1-63057-561-8
ISBN-10: 1-63057-561-5

Printed and bound in the United States of America.

Acknowledgments

Thanks as always to my dad, Paul Tran, for always being there and providing support and honest feedback on all the lessons in the textbook.

I would like to give a special thanks to the team at SDC Publications for their editing and corrections. Thanks also for putting together such a beautiful cover design.

Finally, I would like to thank you, our readers, for your continued support. It is with your consistent feedback that we were able to create the lessons in this book with more detailed and useful information.

Preface

My SOLIDWORKS journey began in my youth, while observing my dad operate his engineering company. As I developed my career, I became more involved in different facets of the software, from design work to project management. After nearly two decades of exposure to SOLIDWORKS, I found myself working through distinctive projects, where hundreds of components are considered small assemblies.

Throughout this time I've had the opportunity to focus on reoccurring challenges and develop unique solutions. One such challenge I faced on a daily basis was the need to quickly create efficient models to print for testing. In a deadline driven work environment, it is crucial to create accurate and manufacturable designs.

The lessons in this manual reflect much of what I do at work. I wanted to share the techniques I've developed to create various models using surfaces. I hope you will find this manual useful and that you can apply these lessons to grow your own skills.

Class Files
To download the files that are required for this book, type the following in the address bar of your web browser:

SDCpublications.com/downloads/978-1-63057-561-8

TABLE OF CONTENTS

Chapter 1: Introduction to Surfaces

Solids or Surfaces – What is the difference

1. Solids vs. Surfaces:

In SOLIDWORKS, both solids and surfaces are known as Solid Bodies and Surface Bodies. They are advanced tools used to easily create complex geometry. These two entities are very similar and may even be considered the same.

Solid Bodies and Surface Bodies are made up of two types of entities: Geometry and Topology. Geometry represents shapes such as a surface, curve, line, arc, spline, or a point. Topology describes relationship such as a face, an edge, or a vertex, which connects or forms an opened or closed boundary.

The images below display two different objects that appear to be different at first glance. Geometrically they are not quite the same. However, topologically. they are alike. Both objects have eight vertices, six faces, and twelve edges.

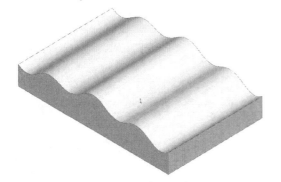
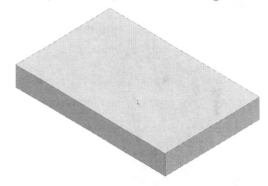

SOLIDWORKS creates surface bodies by forming a mesh of curves called U-V curves. The U curves run along a 4-sided surface and are shown in the magenta color. The curves that run perpendicular to the U curves are called the V Curves, and they are shown in the green color.

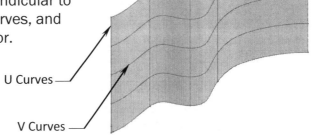

U Curves

Certain commands may offer the preview mesh where these curves are also displayed.

V Curves

2. Patching with continuity:

Blending issues may often arise when creating or patching with surfaces. The three examples below showcase the common examples of patches frequently found in design work.

Contact: (also called CO Continuity)

The yellow surface is in contact with the gray surface. They both share a common edge.

Tangent: (also called C1 Continuity)

The yellow surface is in contact with the gray surface. They share a common edge and are also tangent to each other.

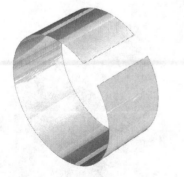

Curvature Continuous: (also called C2 Curvature)

The yellow surface is in contact with the gray surface. They are tangent and share the same curvature.

Each of the blend types has its own features to assist designers in creating the right patch for a surface model. Though every model may have unique needs, these options will provide a smooth transition between two surfaces.

3. When to use surfaces:

Although surface models can be more difficult to create than solid models, solid models may not be enough for all designs. This is where surfacing becomes useful with the important tools to fill in those gaps.

When determining whether to use solid features or surface features, it is helpful to remember that shapes which can be easily made with the Extrude, Sweep, or Loft commands should be made as solid features. These shapes will frequently require more than one side to be created at the same time while residing within the same feature.

A surface model, on the other hand, is created by constructing one surface at a time. This way, different methods and techniques may be used to create the surfaces, one side at at a time. The surface model shown on the right is a primary example of an instance where surfaces should be used.

Surfaces can be used as reference geometry in order to assist with the construction of a model. The surfaces will then be moved into a Surface Bodies folder and become hidden when the model is complete.

4. When not to use surfaces:

Shapes that can be created as one feature along one direction, such as blocks and cylinders, should be made as solid features.

Solid features take less time to make. Generally, the final design of a model should be a solid model. This is because a solid model can have a material assigned to it, can be analyzed to simulate future issues, and can be used to create production drawings.

It is a best practice to combine both solid and surfacing tools (Hybrid Modeling) in order to take advantage of all the tools available and use SOLIDWORKS' powerful features to their fullest extent.

5. The Spline handles:

When creating the surfaces, the Spline tool is often used to create complex curves. This is because it can produce a higher quality curve than other available tools.

Spline can be modified using several controls, including spline points, spline handles, and control polygons. A single spline can have multiple through points and spans (the region between through points). Further, curvature constraints can be applied at each endpoint. At each through point, the tangency vector may be weighed and the tangency direction controlled.

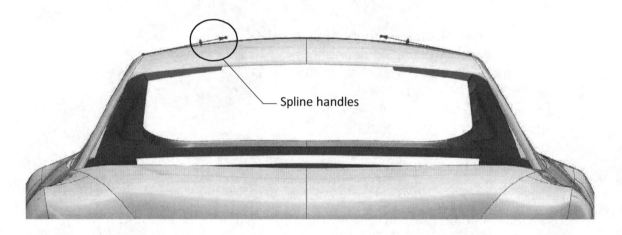

Spline handles

Each Spline includes a set of tools to assist with the manipulation of its curvature. These tools are:

* Curvature Combs
* Spline Handles
* Inflection Points
* Control Polygons

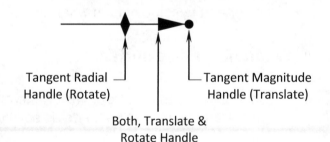

Tangent Radial Handle (Rotate)

Tangent Magnitude Handle (Translate)

Both, Translate & Rotate Handle

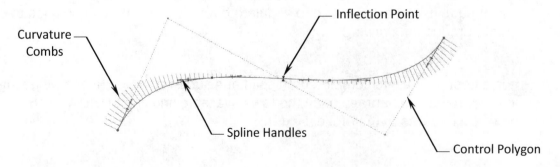

Curvature Combs

Inflection Point

Spline Handles

Control Polygon

6. Check your surface model frequently:

Unlike solid features in which one feature may have multiple sides (i.e. an extruded block has six sides, extruded in one direction), surface models are created one face at a time in order to complete each side of the model. Further, surface models are not frequently verified like solid features.

These surfaces may have small gaps or overlapping elements that are difficult to spot. Therefore, it is important to check your surface model for errors after each feature is built.

The Check (or Check Entity) command, under the Evaluate tab, functions as a tool to check surface models and identify geometry issues. If gaps or overlaps exist in the model, the system will prevent the surfaces from knitting into a solid model. It is a best practice to check your model often to reveal any errors.

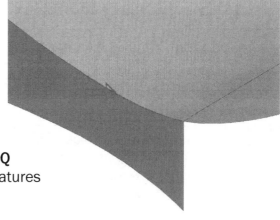

Shortcuts (or hotkeys) are often used to regenerate the surface models.

The hotkeys **Control+B** will rebuild only the new features and their related children, while **Control+Q** will force a complete rebuild of all features in the model.

7. The best approach:

Combining solid and surface tools is the best approach to creating your model. Fully utilizing the collaboration of these tools allows for complex geometry and the optimization of different techniques. Techniques such as using a surface to trim a solid, replacing a surface with different surface, or splitting a solid body into multiple solid bodies, can be necessary to finalize a model.

This unique approach is called Hybrid Modeling, which is a preferred method for most advanced SOLIDWORKS users. The approach, along with other techniques, will be discussed throughout this book when creating the surface models covered in these lessons.

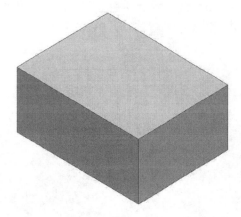

A few important factors to keep in mind from this chapter as we progress through the proceeding lessons:

Solid features are more efficient to use on simple shapes and models. Surface features should be used for complex shapes and models.

Surface features take longer to construct and should be checked frequently for errors.

After all surfaces are created, they should be knitted into one enclosed volume then thickened into a solid model. By turning the surface model into a solid model, we can now analyze for potential issues and create drawings.

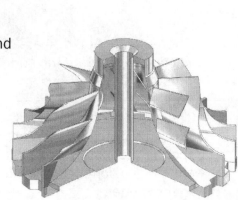

Chapter 2: Surfacing Basics

Extrude, Revolve, Sweep, Loft, and Boundary

This chapter is intended to remind us of the surfacing basics. They consist of five frequently used commands: Extruded Surface, Revolved Surface, Swept Surface, Lofted Surface, and Boundary Surface.

1. Opening a part document:

Open a part document named: **Extruded Surface_Exe.sldprt.**

Switch to the **Surfaces** tab.

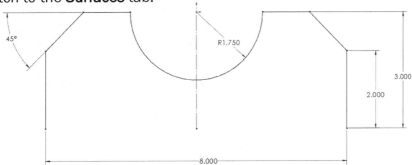

2. Extruding a surface:

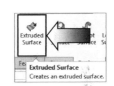

Click **Extruded Surface** (arrow) on the **Surfaces** tab.

For Direction 1, select **Mid Plane**.

For Depth, enter **4.000in.**

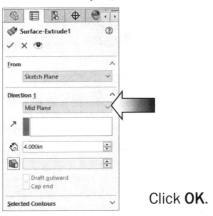

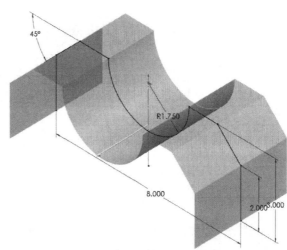

Click **OK.**

3. Creating a Curve Through Reference Point:

This command creates a curve through points or vertices located on one or more planes.

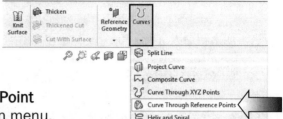

Select the **Curve Through Reference Point** command from the **Curves** drop-down menu.

Select the **2 vertices** as noted.

Click **OK**.

Rotate the surface model towards the back side.

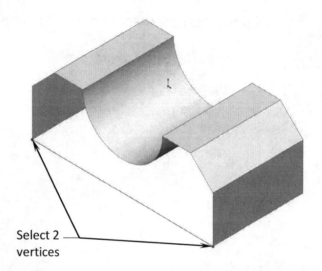

Select 2 vertices

Click **Curve Through Reference Point** again.

Select the **2 vertices** indicated.

Click **OK**.

Select 2 vertices

The front and back sides now have closed boundaries.
They can be closed off with the planar surface command.

4. Creating the planar surfaces:

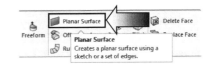

The Planar Surface command creates a flat surface from a set of closed edges.

Click **Planar Surface**.

Select the **8 edges** in the front of the surface model as noted.

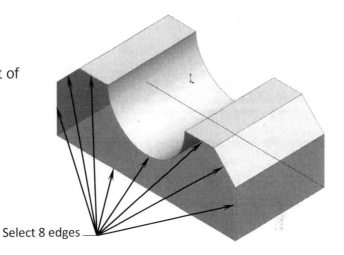

Select 8 edges

The preview graphics display a planar surface is being created from the edges.

Click **OK**.

Click **Planar Surface** again.

Rotate the surface model and select the **8 edges** from the opposite side.

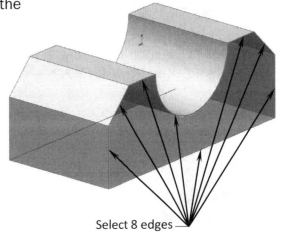

Click **OK**.

Select 8 edges

5. Knitting the surfaces:

The Knit Surfaces command combines two or more surfaces into a single surface.

Click **Knit Surface**.

Select all **3 surface bodies** from the graphics area.

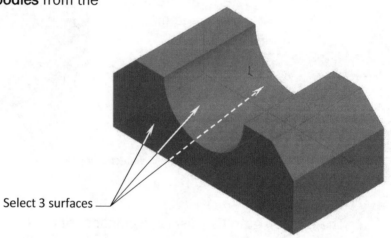

Select 3 surfaces —

Click **OK**.

6. Saving your work:

Select **File, Save As**.

Enter: **Extruded Surface_Completed** for the file name.

Click **Save**.

Close all documents. The next exercise will delve into revolved surfaces.

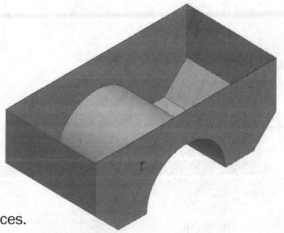

Surfacing Basics – Revolved Surface

This exercise demonstrates the Revolved Surface options. A sketch has been provided to expedite the lesson to focus on the use of the commands.

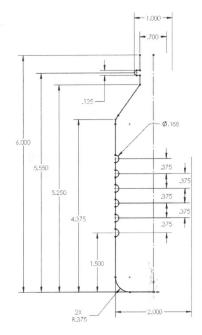

1. Opening a part document:

Open a part document named:
Revolved Surface_Exe.sldprt.

Edit the sketch1; take note of the centerline. This will be used as the Axis of Revolution.

2. Revolving a surface:

Switch to the **Surfaces** tab and click: **Revolved Surface** or select: **Insert, Surface, Revolve.**

The **centerline** is selected automatically as the Axis of Revolution.

For Direction 1, use the default **Blind** type.

For Revolve Angle, use the default **360°**.

Click **OK**.

3. Creating a section view:

Create a section view to quickly look at the interior details.

Click **Section View** on the View Heads-Up tool bar.

Use the default **Front** plane as the Cutting plane.

The preview graphics show a surface model with no thickness.

Push **Esc** to exit the section view command.

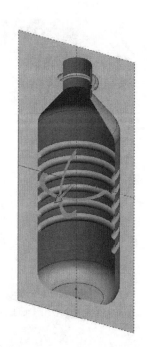

4. Adding fillets:

The sharp edges around the neck and the gripping features must be filleted.

Click **Fillet**.

For Fillet Type, use the default **Constant Size-Radius** option.

For Radius, enter **.060in**.

For Items to Fillet, select **13 edges** as indicated.

Click **OK**.

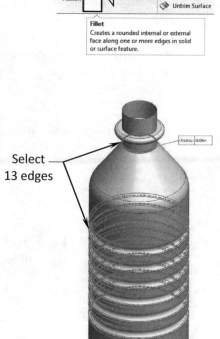

Select 13 edges

The sharp edges are replaced with the .060in. fillets.

5. Saving your work:

Select **Files, Save As**.

Enter **Revolved Surface_Completed.sldprt** for the file name.

Click **Save**.

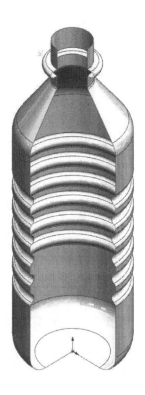

Optional: Partial Section

> _Sketch 2 lines on the Top plane, 90deg apart._

> _Use Trim Surface and remove the center portion of the surface body as shown._

Surfacing Basics – Swept Surface

This exercise demonstrates the use of the Swept Surface command. The Swept Surface command creates a surface by moving a profile along a path. The software creates a series of intermediate sections made by replicating a profile at various positions along the path.

1. Opening a part document:

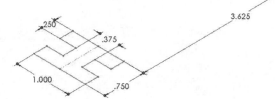

Open a part document named: **Swept Surface_Exe**.

The sweep profile has already been created. We will need to sketch the sweep Path, then sweep the profile along the path to create the main surface body.

2. Sketching the sweep path:

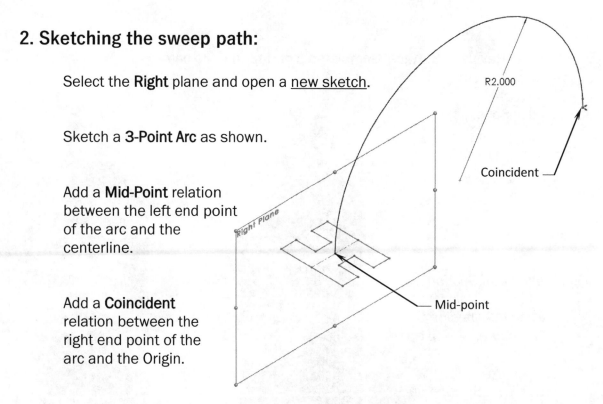

Select the **Right** plane and open a <u>new sketch</u>.

Sketch a **3-Point Arc** as shown.

Add a **Mid-Point** relation between the left end point of the arc and the centerline.

Add a **Coincident** relation between the right end point of the arc and the Origin.

Add the **R2.00in** dimension to fully define the sketch.

<u>Exit</u> the sketch.

3. Creating a swept surface:

Switch to the **Surfaces** tab and click: **Swept Surface** or select: **Insert, Surface, Sweep.**

For Sweep Profile, select one of the entities of the **Sketch1** in the graphics area.

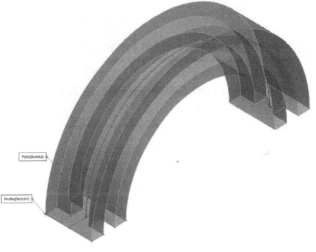

For Sweep Path, select the **Arc** of Sketch2.

Click **OK**.

4. Saving your work:

Select **File, Save As**.

Enter: **Swept Surface_Completed** for the file name.

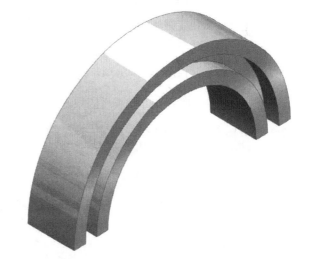

Click **Save**.

Close all documents. The next lesson focuses on the Lofted Surface options.

Surfacing Basics – Lofted Surface

This exercise explores the Lofted Surface options.
The Lofted Surface command creates a surface
by making transitions between profiles.
Guide curves are also used to help eliminate
or control the twists between the profiles.

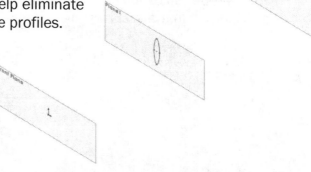

1. Opening a part document:

Open a part document named:
Lofted Surface_Exe.sldprt

You will find two additional planes with existing sketches. This exercise calls
for a total of three sketches. A third sketch will be created on the next plane,
which will be the Front Plane.

2. Sketching the 3rd profile:

Select the **Front** plane and open a
new sketch.

Sketch an **Ellipse** (arrow) centered on the Origin.

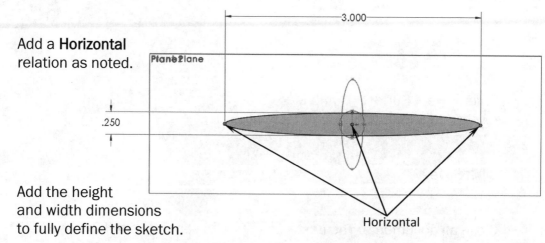

Add a **Horizontal**
relation as noted.

Add the height
and width dimensions
to fully define the sketch.

Exit the sketch.

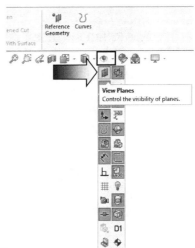

Click-off **View Planes** under the Visibility drop-down list to hide all planes.

3. Creating a lofted surface:

Switch to the **Surfaces** tab and click **Lofted Surface**.

For Loft Profiles, select the **3 profiles** in the order indicated in the image below. (Profile 1 = Left point, Profile 2 = Top point, Profile 3 = Right point.)

There should be a slight twist in this surface model. By selecting the connectors in the order mentioned above, the intentional twists can be created without the need of the guide curves.

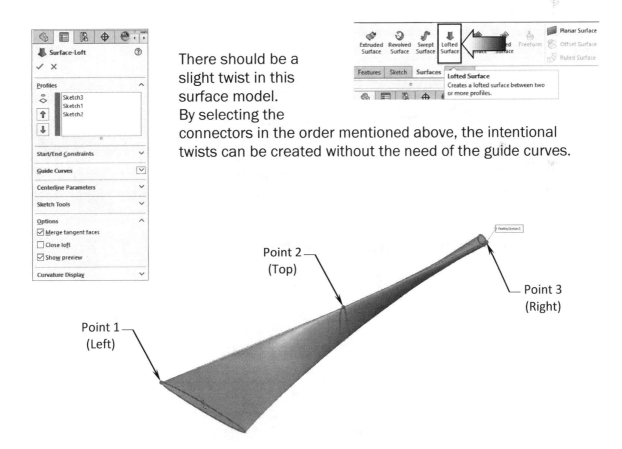

Click **OK**.

The lofted surface body is created using 3 loft profiles, and the twist feature is added into the lofted surface by alternating the connectors in the profiles.

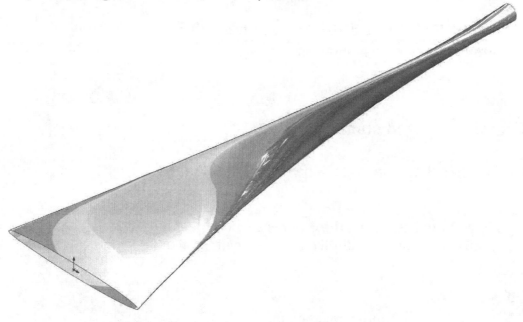

4. Saving your work:

Click **Files, Save As**.

Enter **Lofted Surface_Completed** for the file name.

Click **Save**.

Close all documents.
The next exercise examines the Boundary Surfaces options.

Surfacing Basics – Boundary Surface

In this exercise, we will examine the Boundary Surface tool.
The boundary surface feature allows the creation of surfaces
that can be tangent or curvature continuous on all sides
of the surface. In most cases, this delivers a better
result than the loft tool.

1. Opening a part document:

Open a part document named: **Boundary Surface_Exe.sldprt**

There are 3 sketches created on 3 parallel planes. A set
of 3D Splines will be created to connect the profiles.
The Splines will be alternated to create the slight twists
in the surface model.

2. Sketching the 3D splines:

Click the **Sketch** drop-down and select **3D Sketch**.

Select the **Spline** command
and sketch a **3-point
Spline** in the order
shown on the right.

Point 1
(Right)

A Coincident relation
is added automatically
at the connecting
point.

Point 2
(Front)

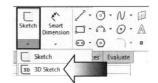

Point 3
(Left)

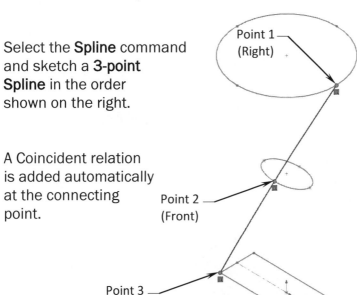

Add two **centerlines** as shown, along the Y direction (vertical).

Add **Tangent** relation between the centerlines and the Splines.

Tangent

Tangent

Do not change the Tangent Weight by moving the Spline Handles. The default tangent weights are good for this exercise.

Sketch another **3-point Spline** in the order shown in the image on the right.

(Point 1 = Front
Point 2 = Left
Point 3 = Left, Rear)

Point 1
(Front)

Point 2
(Left)

Point 3
(Left, Rear)

Next, add two centerlines and two tangent relations to match the tangent weight to the first spline.

Do not adjust the spline handles.

Sketch two additional **3-point Splines** (total of 4) following the same order as the prior two mentioned above.

Change to different view orientations to check your splines against the images shown below.

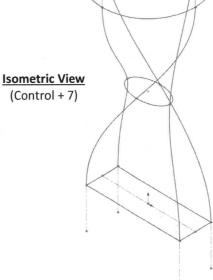

Isometric View
(Control + 7)

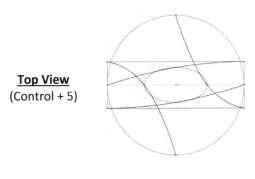

Top View
(Control + 5)

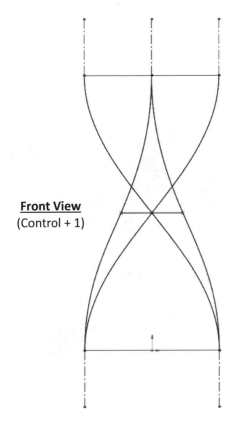

Front View
(Control + 1)

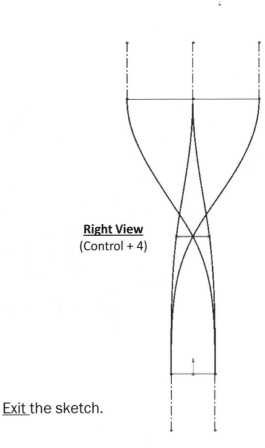

Right View
(Control + 4)

<u>Exit</u> the sketch.

3. Creating a boundary surface:

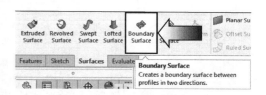

Due to the excellent results, designs that require high quality surfaces with curvature continuity should use this tool.

Click **Boundary Surface** on the **Surfaces** tab.

For Direction 1, select the **3 profiles** as indicated.

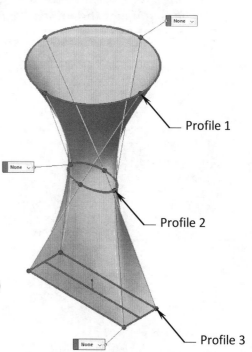

Profile 1

Profile 2

(The Direction 2 splines are hidden for clarity only.)

Profile 3

For Direction 2, select the **1st spline** and click **OK** on the **SelectionManager** pop-up dialog.

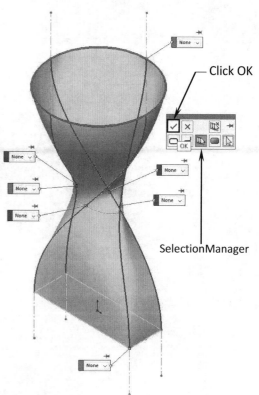

Click OK

Select all **4 splines** and click the Selection-Manager each time. Check your preview against the image shown on the right.

SelectionManager

Click **OK**.

The Boundary Surface is created
using three sketch profiles as direction 1,
and four 3D Sketch splines as direction 2.

4. Saving your work:

Select **File, Save As**.

Enter: **Boundary Surface_Completed**
for the file name.

Click **Save**.

Close all documents.
This concludes the chapter reviewing Surfacing Basics. The following chapters
contain lessons with Advanced Surfacing methods using SOLIDWORKs.

Exercise: Extrude and Trim Surfaces

The exercises give you an opportunity to apply what you have learned from the lesson. The instructions are limited to allow you to try things out using your own techniques, at your own pace.

1. Opening a part document:

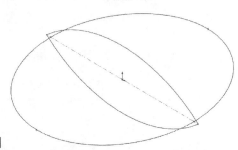

Browse to the Training Folder and open a part document named:
Lesson 2_Exercise.sldprt.

There are 2 sketches in this model. They will be extruded as surfaces and trimmed to the final shape and size.

2. Extruding the 1st surface:

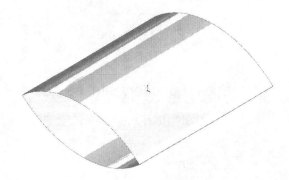

Select **Sketch1** and extrude it using the following parameters:

Midplane

6.00in.

Click **OK**.

3. Extruding the 2nd surface:

Select **Sketch2** and extrude it using the following parameters:

Midplane

3.00in.

Click **OK**.

4. Extruding the 2ⁿᵈ surface:

Create a **Trimmed Surface** using the following:

Type: **Mutual**

Trim Surfaces:
Surface-Extr1
Surface-Extr2

Keep Selections:
3 Pieces (as noted)

Click **OK**.

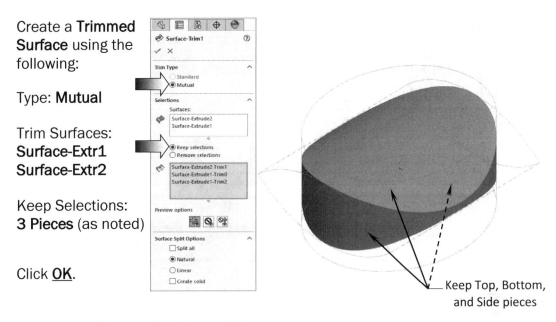

Keep Top, Bottom, and Side pieces

5. Adding the 1ˢᵗ Variable Size fillet:

The Fillet tool on the Surface tab is the same as the one on the Features tab.

Click **Fillet**.

For Fillet Type select **Variable Size Fillet**.

Click the upper edge of the model.

Four Control Points are added and placed 25% apart.

Enter the radius values as shown.

Click **OK**.

.250in

.625in

.625in

.250in

6. Adding the 2nd Variable Size fillet:

Click **Fillet** again.

For Fillet Type select **Variable Size Fillet**.

Click the <u>lower edge</u> of the model.

Four Control Points are added.

Enter the radius values as shown.

Click **OK**.

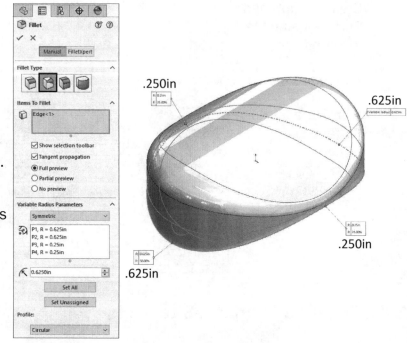

7. Saving your work:

Select **File, Save As**.

Enter **Lesson 2_Exercise_Completed** for the file name.

Click **Save**.

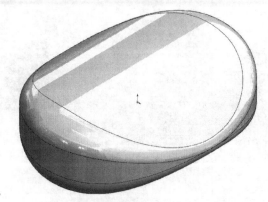

Close all documents.

Chapter 3: Using Boundary and Lofted Surface
Shoehorn

In this lesson, we will build a model of a shoehorn in order to explore a variety of frequently used commands when creating surfaces. Furthermore, we will go through the necessary steps to convert a surface model into a solid model.

1. Starting a new part document:

Select **File**, **New Part** and select the **Part** template.

Set the Units to **Inches** and the number of decimals to **3**.

Select the <u>Right</u> plane and open a **new sketch**.

Sketch a **Circle** and a couple of **centerlines** as shown below.

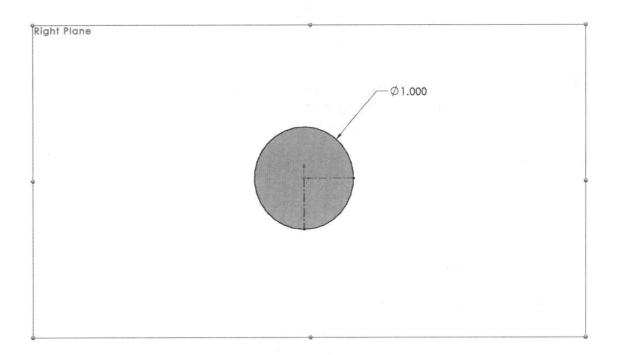

Add a diameter dimension to fully define the circle.

Sketch a 2-point **Spline** and trim away the lower left portion of the circle as shown in the image.

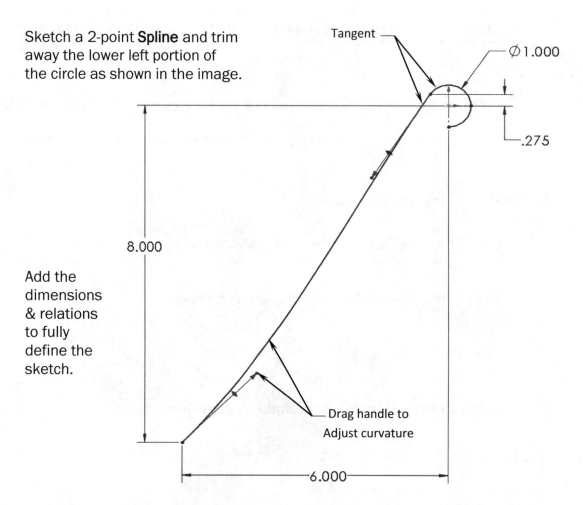

Add the dimensions & relations to fully define the sketch.

This lesson focuses specifically on the use of surfacing tools. Therefore, the curvature of the spline is not required to be accurate.

Optional: To modify the lower spline handle, add one or more construction lines and reference dimensions, then move the handle towards the corner as needed.

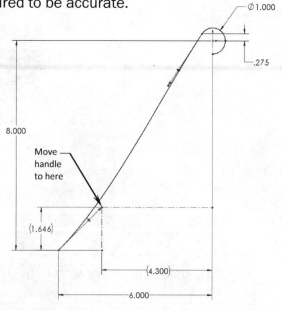

The upper spline handle does not need to be modified because it is constrained by the tangent relation.

<u>Exit</u> the sketch.

2. Creating the 1st reference plane:

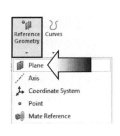

In order to create the two curves on both ends of the previous sketch, a new plane must be created.

Click **Plane** from the **Reference Geometry** drop-down menu.

Select the **Spline** for the First Reference.

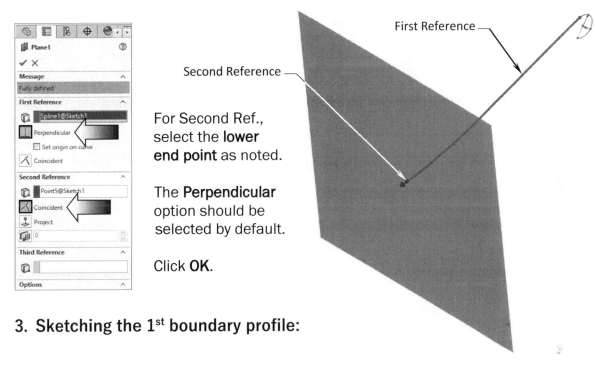

First Reference

Second Reference

For Second Ref., select the **lower end point** as noted.

The **Perpendicular** option should be selected by default.

Click **OK**.

3. Sketching the 1st boundary profile:

Open a **new sketch** on Plane1.

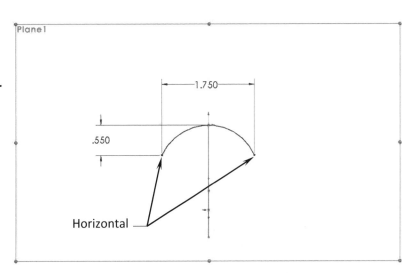

Sketch a **3-Point Arc**, then add the relations as noted.

To fully define the sketch, add the dimensions and relation as noted.

Exit the sketch.

Plane1

1.750

.550

Horizontal

The 1st boundary profile is created at the lower end of the path. The 2nd profile will be sketched on the upper end.

4. Sketching the 2nd boundary profile:

Select the <u>Front</u> plane and open a **new sketch**.

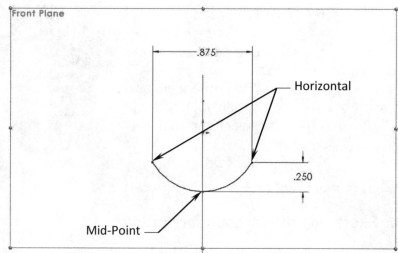

1st boundary profile

Sketch another **3-Point Arc** and add the relations as indicated.

Add the vertical and horizontal dimensions to fully define the sketch.

2nd boundary profile

<u>Exit</u> the sketch.

The 2nd boundary profile is created on the upper end.

5. Comparing methods:

Outlined below are several methods which may be used to create the main surface body. Though there are many ways to achieve this design, the boundary function often produces the most accurate results.

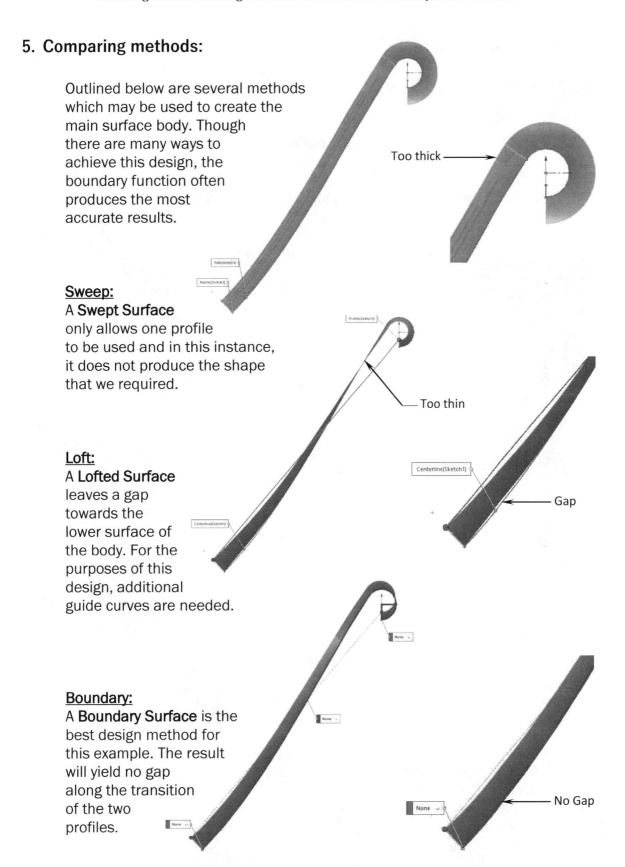

Too thick

Sweep:
A **Swept Surface** only allows one profile to be used and in this instance, it does not produce the shape that we required.

Too thin

Loft:
A **Lofted Surface** leaves a gap towards the lower surface of the body. For the purposes of this design, additional guide curves are needed.

Gap

Boundary:
A **Boundary Surface** is the best design method for this example. The result will yield no gap along the transition of the two profiles.

No Gap

6. Creating a boundary surface:

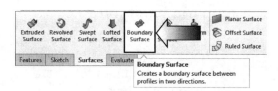

Switch to the **Surfaces** tab.

Click **Boundary Surface** on the **Surfaces** tab.

For Direction 1, select the **2 boundary profile** sketches.

For Direction 2, select the sketch of the **Spline**.

Keep all other parameters at their default values.

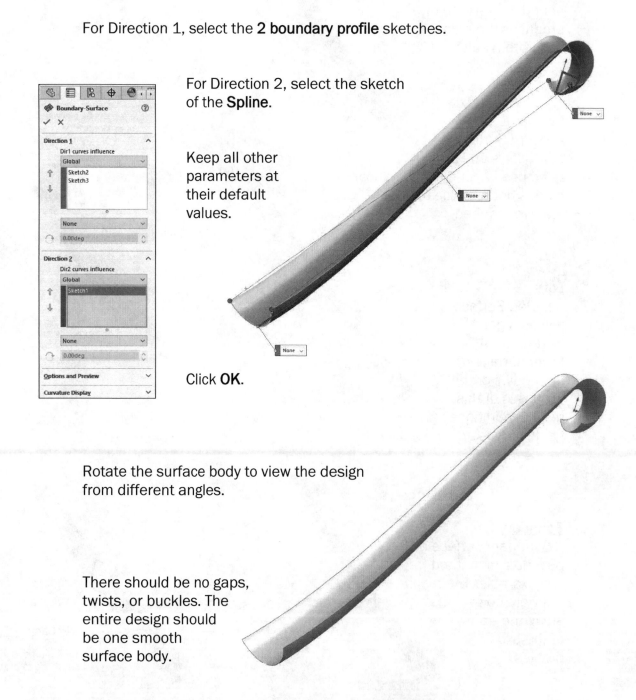

Click **OK**.

Rotate the surface body to view the design from different angles.

There should be no gaps, twists, or buckles. The entire design should be one smooth surface body.

7. Creating the 2nd reference plane:

Over the next few steps we will create the trimming profiles in order to round off the two ends. Because the surface body was created at an oblique angle, a new plane is required once again.

Click **Plane** from the **Reference Geometry** drop-down list.

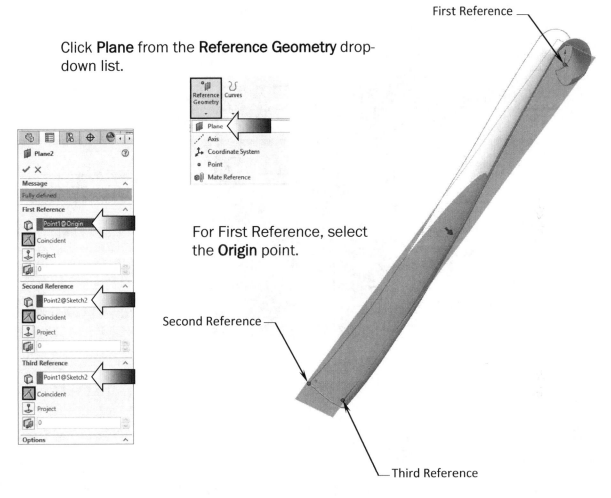

First Reference

For First Reference, select the **Origin** point.

Second Reference

Third Reference

For Second Reference, select the **Vertex** on the <u>lower left side</u> of the surface body.

For Third Reference, select the **Vertex** on the <u>lower right side</u> of the surface body.

Click **OK**.

8. Constructing the 1st trim sketch:

Select the new <u>Plane2</u> and open a **new sketch**.

Sketch a **3-Point Arc** and add the relations shown in the image below.

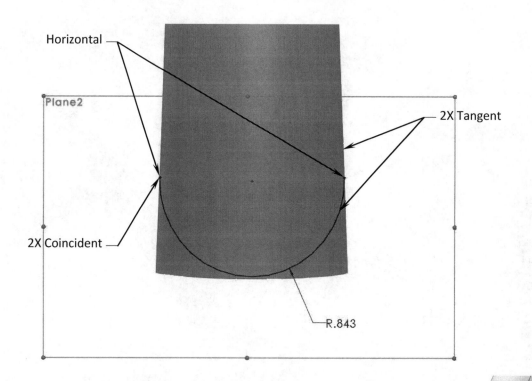

Horizontal

Plane2

2X Tangent

2X Coincident

R.843

Add the radius dimension to fully define the sketch.

Because "Extruded Cut" is not a featured option in the surfaces toolbar, this sketch will be utilized as the trim tool for the dual purposes of removing the lower portion of the surface body, while creating the rounded end.

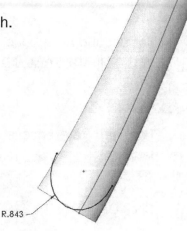

R.843

9. Trimming the bottom:

The Standard Trim option is automatically selected when a sketch is used as the trim tool.

Click **Trim Surface**.

Delete Face	Extend Surface		Thicken
Replace Face	Trim Surface		Thickened C
	Un...		

Trim Surface (T)
Trims a surface where one surface intersects with another surface, a plane, or a sketch.

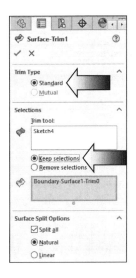

The Trim Type should be defaulted to **Standard Trim**.

For Trim Tool, the **Sketch4** (3-Point Arc) should be selected already, if not, select the 3-Point arc.

Click the **Keep Selections** option (arrow).

Keep Selections

For Keep Selections, select the **inner portion** (larger) of the surface body as indicated.

Click **OK**.

Inspect your surface model against the image shown here.

10. Constructing a split sketch:

In the steps below, we will utilize another method to round off the top end of the surface body. **Split Line** is an option that is used quite frequently in advanced surface modeling.

The Split Line command uses a sketch to project and split a surface into two or more surface bodies. The new surfaces can be moved, rotated, copied, or deleted.

Select Plane2 from the FeatureManager tree and open a **new sketch**.

Sketch a **3-Point Arc** and add the relations indicated in the image.

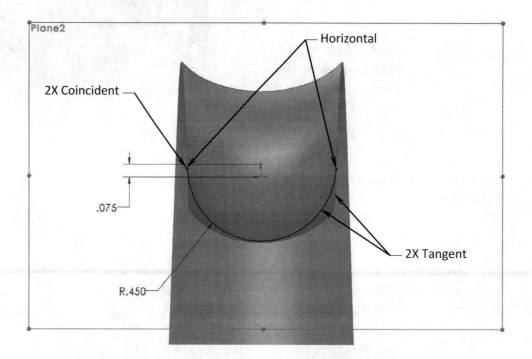

Add dimensions to fully define the sketch. Notice the arc is drawn slightly above the edge of the surface model.

Remain in sketch mode for the next step.

11. Creating a split line feature:

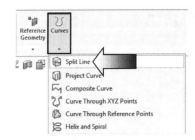

Click **Split Line** from the **Curves** drop-down menu.

For Type of Split, select the **Projection** type.

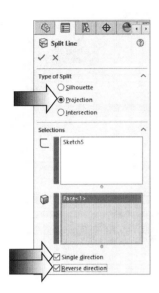

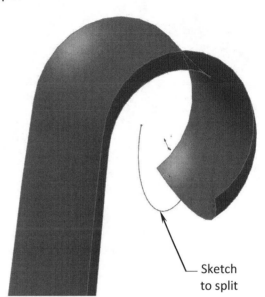

Sketch
to split

For Sketch to Project, **Sketch5** should be selected by default. If not, select it
either from the FeatureManager tree or directly
from the graphics area.

For Faces to Split, select the **Surface
Body** in the graphics area.

Enable the 2 checkboxes:
* **Single Direction**
* **Reverse Direction**

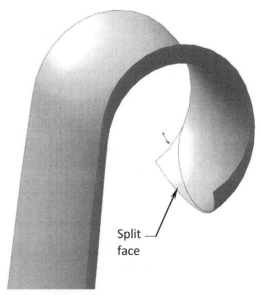

Split
face

Click **OK**.

12. Deleting faces:

Click **Delete Face**.

For Faces to Delete, select the **lower portion** of the surface body as noted.

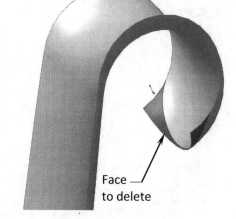

Face to delete

Under Options, click **Delete** (arrow).

Click **OK**.

The lower portion of the surface body is removed, leaving a nicely rounded shape along the perimeter of the model.

Explanation of options:

Delete Face: Deletes one or more faces from a surface body.

Delete and Patch: Deletes one or more faces and automatically patches and trims the body.

Delete and Fill: Deletes and generates a single face to close any gap.

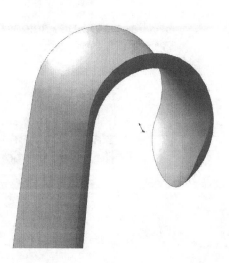

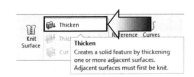

13. Thickening the surface model:

Now that we have completed the end details, the surface model is now ready to be thickened.

Click **Thicken** on the **Surfaces** tab.

For Surface to Thicken, select the **surface body** from the graphics area as noted.

For thickness location, select **Both Sides**.

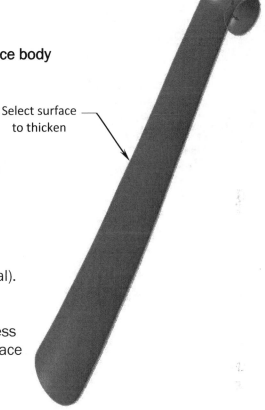

Select surface to thicken

For Thickness, enter **.035in** (.070in total).

Zoom in closer to verify that the thickness is being added to both sides of the surface body.

Click **OK**.

.035in per side, .070in total thickness

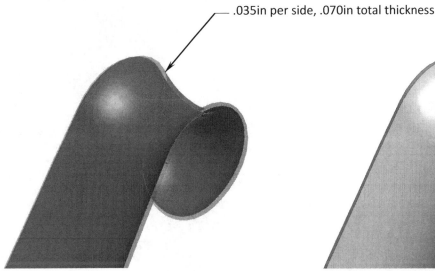

14. Adding fillets:

The Full Round Radius option automatically creates fillets that are tangent to three adjacent face sets. The Left and Right Face Sets can be selected in any order, but the Center Face Set must be the ones that connect with the Left and Right faces.

Click **Fillet** and select **Full Round** option.

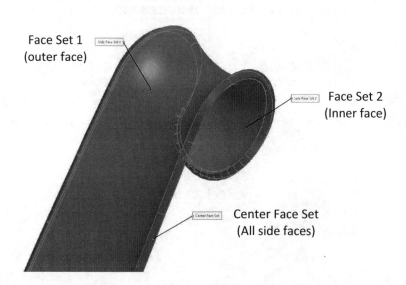

Face Set 1
(outer face)

Face Set 2
(Inner face)

Center Face Set
(All side faces)

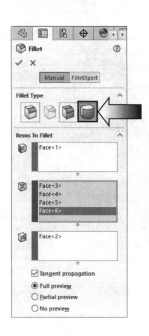

For Face Set 1, select the face on the **left side** of the model.

For Center Face Set, right-click one of the side faces and choose: **Select Tangency** (be sure to select <u>all side faces</u>).

For Face Set 2, select the face on the **right side** of the model as noted.

Click **OK**.

15. Saving your work:

Click **File, Save As**.

Enter **Shoehorn_Completed.sldprt** for the file name.

Click **Save**.
(Keep the model open.)

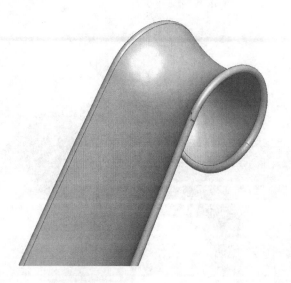

16. RealView Graphics:

Presentation is an important aspect of the design. Now that our model is completed, we will use RealView and Ambient Occlusion to incorporate realistic details without the need of rendering. PhotoView360 can be used within SOLIDWORKS, and alternatively, Visualize, a standalone suite of tools, can be used to produce photo-quality renderings of any model.

RealView Graphics is hardware (graphics card) support of advanced shading in real time, including self-shadowing and scene reflections. To apply an appearance use RealView Graphics for a realistic model display.

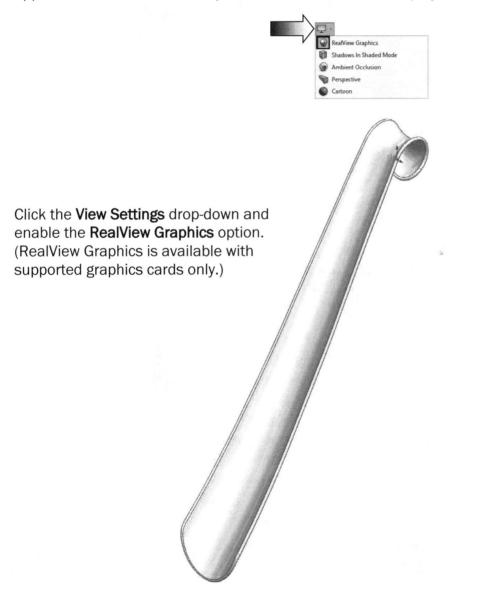

Click the **View Settings** drop-down and enable the **RealView Graphics** option. (RealView Graphics is available with supported graphics cards only.)

17. Ambient Occlusion:

Ambient occlusion is a global lighting method that adds realism to models by controlling the attenuation of ambient light due to occluded areas. Objects appear as they would on an overcast day.

To change the display quality level for Ambient Occlusion, click **Tools, Options**. On the **System Options** tab, click **Display**, then select or <u>clear</u> **Display draft quality ambient occlusion**.

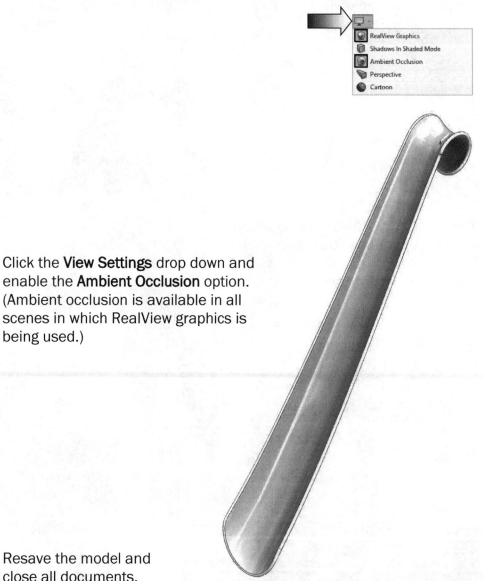

Click the **View Settings** drop down and enable the **Ambient Occlusion** option. (Ambient occlusion is available in all scenes in which RealView graphics is being used.)

Resave the model and close all documents.

Using Surface Trim & Lofted Surface

In the following steps, the sketch has been pre-designed in order to keep the focus of this lesson on the **Trim Surface** and **Lofted Surface** commands.

18. Opening a part document:

Open a part document named:
Surface Trim_Loft.sldprt

There are 3 sketches on the FeatureManager tree.
Sketch1 will be used to create the main surface body.

19. Creating a revolved surface:

Select **Sketch1** on the FeatureManager tree and click **Revolved Surface**.

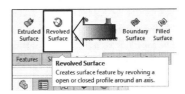

The **centerline** should be selected automatically as the Axis of Revolution.

For Direction 1, select the **Mid Plane** option from the drop-down list.

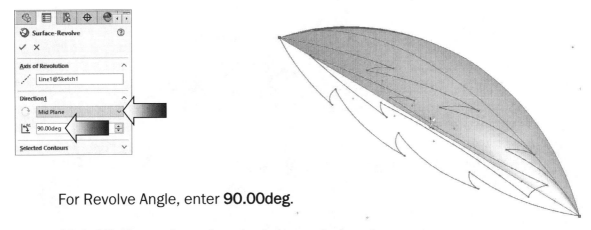

For Revolve Angle, enter **90.00deg.**

Click **OK**. The main surface body is created.

20. Trimming with a sketch:

The Standard Trim Type is used when trimming with a sketch.

Click **Trim Surface**.

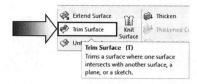

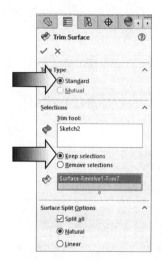

For Trim Type, select the **Standard** option.

Keep Selections
(Select the inside)

For Trim Tool, select **Sketch2** from the Feature Manager tree.

For Keep Selections, select the **inner portion** of the main surface body as noted.

Click **OK**.

21. Making a surface offset:

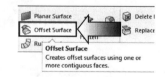

A copy of the main surface needs to be made and trimmed to shape in the next couple steps.

Click **Offset Surface**.

For Offset Parameters, select the **main surface body**.

For Offset Distance enter **.050in**.

Click **Reverse** to place the copy <u>above</u> the main surface body.

Click **OK**.

22. Trimming with another sketch:

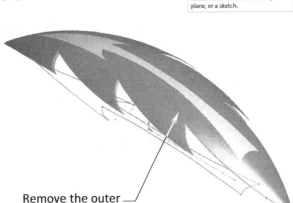

The arrow feature in the middle of the main surface body will be created using Sketch3.

Click **Trim Surface**.

For Trim Type, click **Standard**.

Remove the outer portion of the offset surface

For Trim Tool, select **Sketch3**.

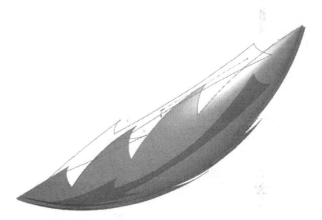

For Remove Selections, select the <u>outer</u> portion of the **Offset Surface** body and the <u>inner</u> portion of the **Main Surface** body as noted.

Under Split Options, enable the options shown in the dialog box.

.050" gap

Click **OK**.

There is a .050" gap between the 2 surfaces. We will fill this gap with the lofted surface tool in the next step.

23. Creating a lofted surface:

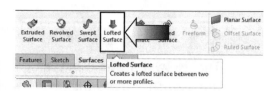

Though there are several options which may be used to fill the gap between the two surfaces, in this example, we will use the lofted surface method.

Click **Lofted Surface**.

Right-click in the graphics area and click: **SelectionManager** (arrow).

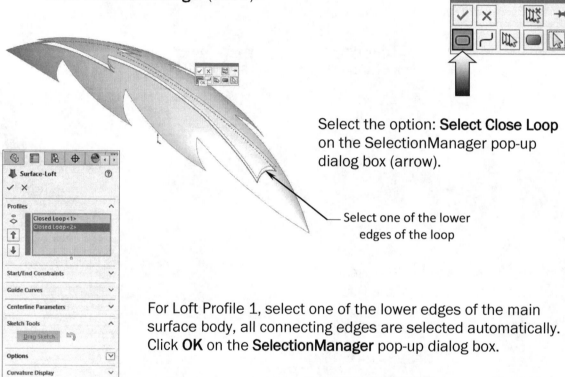

Select the option: **Select Close Loop** on the SelectionManager pop-up dialog box (arrow).

Select one of the lower edges of the loop

For Loft Profile 1, select one of the lower edges of the main surface body, all connecting edges are selected automatically. Click **OK** on the **SelectionManager** pop-up dialog box.

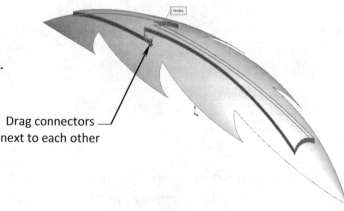

For Loft Profile 2, select all the lower and **inside edges** of the main surface. Drag the connectors so they are positioned right on top of each other.

Drag connectors next to each other

Click **OK**.

24. Knitting all surfaces:

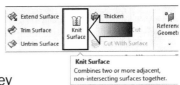

The Knit Surface command joins the surfaces and heals small gaps in the surface model as long as they are within the knitting tolerance. In this case the gaps are less than .0005", which can be healed fairly easily.

Click **Knit Surface**.

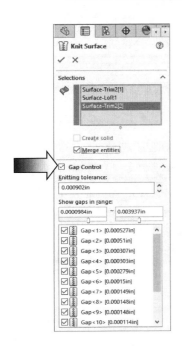

Select 3 surfaces

For Selections, select all **3 surfaces** either from the graphics area or on the FeatureManager tree.

Enable the **Gap Control** checkbox and click one of the **Gap checkboxes**, all other checkboxes will be selected automatically.

Enable the **Merge Entities** option and keep the Knitting Tolerance at its default value.

Click **OK**.

25. Saving your work:

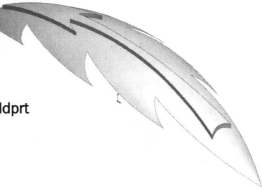

Select **File, Save As**.

Enter: **Surface Trim_Loft_Completed.sldprt** for the file name.

Click **Save**.

<u>*Optional:*</u>

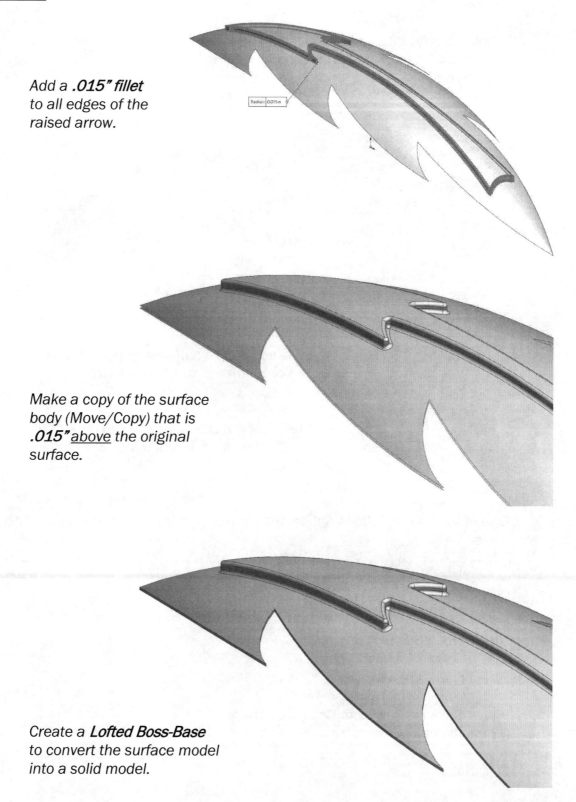

Add a *.015" fillet*
to all edges of the
raised arrow.

Make a copy of the surface
body (Move/Copy) that is
.015" <u>above</u> the original
surface.

Create a **Lofted Boss-Base**
to convert the surface model
into a solid model.

Exercise: Creating a Lofted Surface

The exercises give you an opportunity to apply what you have learned from the lesson. The instructions are limited to allow you to try things out using your own techniques, at your own pace.

1. Opening a part document:

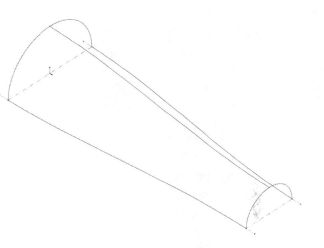

Browse to the Training Folder and open a part document named: **Lesson 3_Exercise.sldprt**.

There are 2 Sketch Profiles and 3 Guide Curves in this model. They will be used to create a Surface body using the Loft tool.

There are also 4 construction lines in the Side Guide Curves sketch. The splines were constrained tangent to these construction lines.

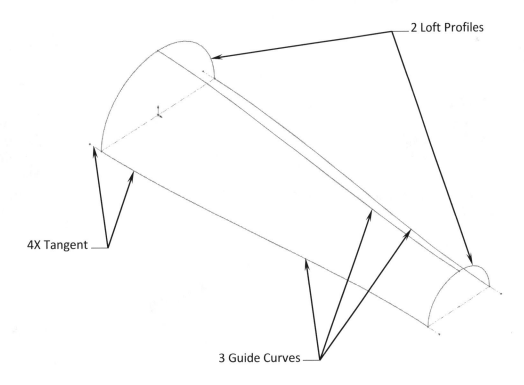

2 Loft Profiles

4X Tangent

3 Guide Curves

2. Creating a surface loft:

Click **Lofted Surface**.

For Loft Profiles
select **Profile1**
and **Profile2**
either from the
Feature tree or
directly in the
graphics area.

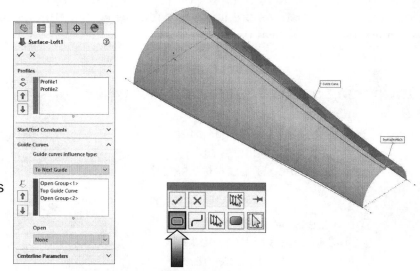

For Guide Curves
select the **3
guide curves**
in the graphics
area.

Click the **green check mark** on the **SelectionManager** pop-up dialog when
selecting the splines each time. This step is needed since the splines were
created together in the same sketch.

Click **OK**.

3. Saving your work:

Select **File, Save As**.

Enter **Lesson 3_Exercise_Completed**
for the file name.

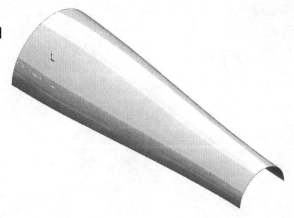

Click **Save**.

Close all documents.

Chapter 4: Multibody Designs
Using Surface Trim, Offset, and Loft

This chapter will teach us the use of Multibody Design using SOLIDWORKS where an existing body is used as a template to help define the size, shape, and location of the other bodies.

1. Opening a part document:

Select **File, Open**.

Locate a part document named: **Bottle with Cap.sldprt** from the Training Files folder and open it.

There are 2 sketches already created on the Feature tree.

2. Creating a revolved surface:

Switch to the **Surfaces** tab.

Select **Sketch5** from the Feature tree and click **Revolved Surface**.

For Axis of Revolution, select the **vertical centerline** as noted.

For Direction 1, use the default **Blind** option and **360°**.

Click **OK**.

Revolve Axis

3. Creating a trimmed sketch:

Select the <u>Right</u> plane and open a **new sketch**.

Right-click one of the edges as noted and pick: **Select-Tangency**.

Click **Offset Entities** and enter **.010** for Distance.

The offset is on the <u>outside</u>.

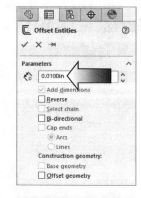

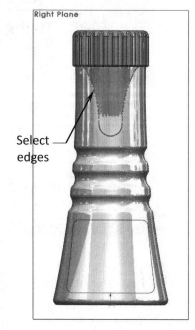

Select edges

Outside ⟶

Add line and trim

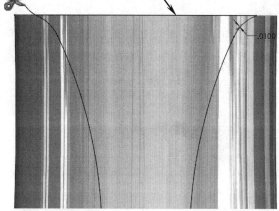

(The Cap body is hidden for clarity.)

Add a **Line** across the top and **trim** away the entities as noted.

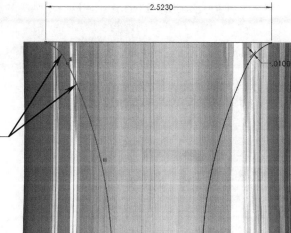

Add a **Tangent** relation to the 2 curves as indicated.

Tangent

Add the **2.523** dimension to fully define the sketch.

Switch to the **Surfaces** tab.

Click **Trim Surface**.

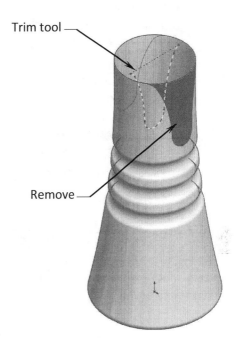

Trim tool

Remove

For Trim Type,
select **Standard**.

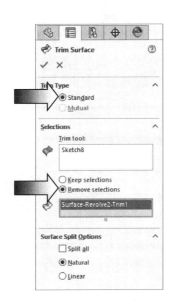

For Trim Tool,
select **Sketch8**
from the Feature-
Manager tree.

Select **Remove Selections**
and click <u>inside</u> the area
as indicated in the callout
to remove this portion.

Click **OK**.

Inspect your model
against the image
shown here.

4. Adding thickness:

Click **Thicken**.

For Thicken Parameters, select the **Surface body** from the graphics area.

For Thickness Direction, select **Inside**.

For Thickness, enter **.060in**.

Click **OK**. (The model is thickened into a solid model.)

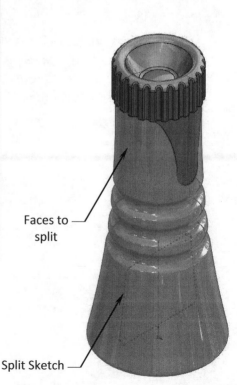

5. Creating a split line feature:

Click **Curves, Split Line**. (The Cap body is shown.)

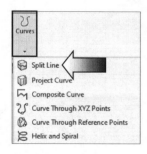

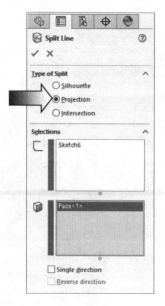

For Type of Split, select **Projection**.

For Split Sketch, select **Sketch6** from the Feature tree.

For Faces to Split, select the **Surface Body** from the graphics area.

Click **OK**.

Faces to split

Split Sketch

6. Making an offset surface:

This step creates a raised surface for the label area.

Click **Offset Surface**.

For Offset Parameters, select the **surface** as noted.

For Offset Distance, enter: **.030in**.

The offset surface is on the <u>outside</u>.

Click **OK**.

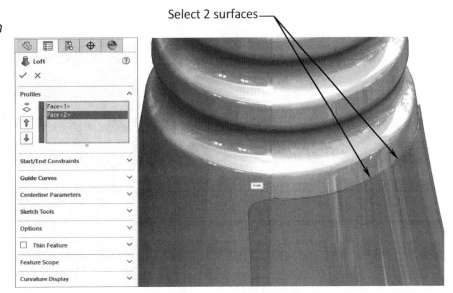

Outside

7. Creating a raised label area:

Switch to the **Features** tab.

Click **Lofted Boss-Base**.

For Loft Profiles, select the **Split Surface** and the **Offset Surface** in the graphics area (<u>Note</u>: drag the two connectors next to each other, if needed).

Leave all parameters at their default values.

Select 2 surfaces

Click **OK**.

8. Creating a 2ND raised feature:

Repeat step number 7 to create another raised feature on the opposite side of the model.

<u>NOTE:</u> *The Cap is Hidden and the two raised features are changed to different color for clarity only.*

9. Adding a new plane:

Click **Reference Geometry, Plane.**

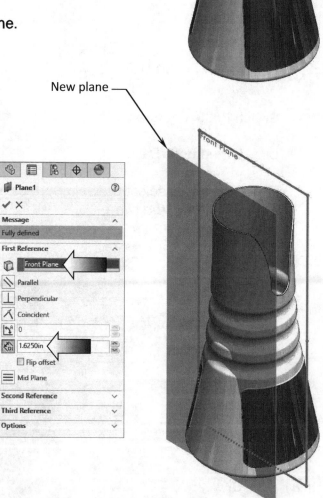

New plane

For First Reference, select the **Right** plane from the FeatureManager tree.

For Offset Distance, enter **1.625in.**

The new plane should be placed on the <u>left side</u> of the model.

Click **OK.**

10. Creating a boss feature:

Select the new <u>Plane1</u> and open a **new sketch**.

Sketch a **Circle** and add the dimensions shown.

Add a **Vertical** relation to fully define the sketch.

Switch to the **Features** tab and click **Extruded Boss-Base**.

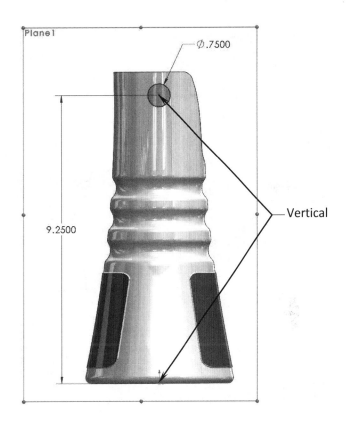

Click <u>Reverse</u> and select the **Up to Next** option.

Click the <u>Draft</u> button and enter **45°** for Draft Angle, also enable the **Draft-Outward** checkbox.

Expand the <u>Feature Scope</u> section and select the **Bottle** from the graphics area, to merge this feature to it.

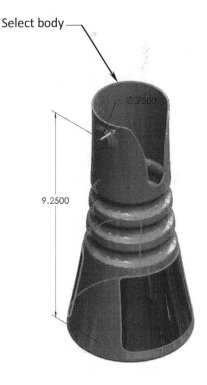

Click **OK**.

11. Adding a .250" fillet:

Click **Fillet**.

Use the default **Constant Size Fillet** option.

For Items to Fillet, select the **edge** as noted.

For Radius, enter **.250in**.

Click **OK**.

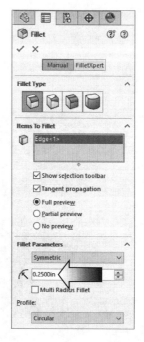

Select edge

12. Adding a .060" fillet:

Click **Fillet** again.

Use the default **Constant Size Fillet** option.

For Items to Fillet, select the **edge** as noted.

For Radius, enter **.060in**.

Click **OK**.

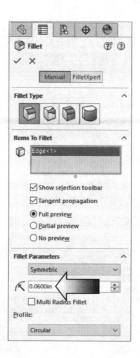

Select edge

13. Adding a pin hole:

Open a **new sketch** on the <u>face</u> indicated.

Concentric ————   ⌀.385

Sketch a new **Circle** and constrain it with a diameter dimension and a concentric relation.

Switch to the **Features** tab.

Sketch face ————

Click **Extruded Cut**.

For Direction 1 use the **Blind** option.

Select body ————

For Depth, enter: **.125in**.

For Feature Scope, select the body of the **bottle** from the graphics area. (This option allows only the Bottle to have the cut, not the other bodies.)

Click **OK**.

14. Mirroring features:

Select **Mirror** from the **Features** tab.

For Mirror Face/Plane select the **Front** plane from the Feature tree.

For Features to Mirror select the **Boss-Extrude1** the **Fillet4**, **Fillet5**, and the **Cut-Extrude2** from the Feature Tree.

Click **OK**.

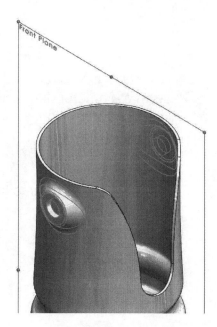

15. Creating the handle sketch:

Open a **new sketch** on the <u>Right</u> plane.

Sketch the profile of the handle as shown.

Either use the **Sketch Mirror** tool or the **Symmetric** relations to keep the profile symmetrical.

Add the dimensions shown to fully define the sketch.

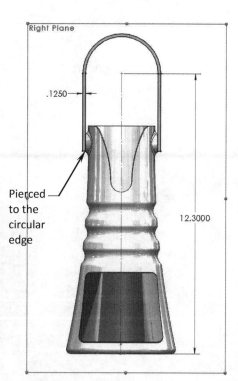

16. Extruding the handle:

Switch to the **Features** tab.

Click **Extruded Boss-Base**.

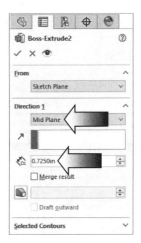

For Direction 1 select the **Mid-Plane** option.

For Depth, enter **.725in**.

Click **OK**.

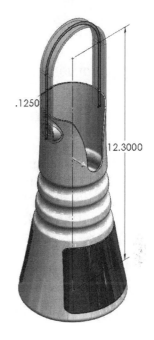

17. Adding Fillets:

Click **Fillet**.

For Fillet Type, use the default **Constant Size Fillet**.

For Items to Fillet, select the <u>4 edges</u> at the bottom of the handle as indicated.

For Radius, enter **.362in**.

Click **OK**.

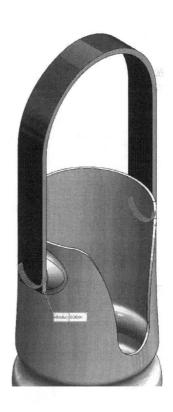

18. Adding a hole:

Open a **new sketch** on the <u>side face</u> of of the handle.

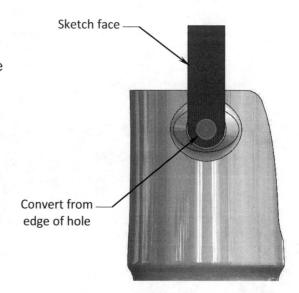

Sketch face

Convert the circular edge of the hole previously created in step 13.

Convert from edge of hole

An **On-Edge** relation is created for the converted circle. If the hole diameter is changed, the converted circle will also change.

Switch to the **Features** tab.

Click **Extruded Cut.**

For Direction 1, select **Through All.**

Click **Reverse Direction.**

For Feature Scope, select the **Handle** from the graphics area.

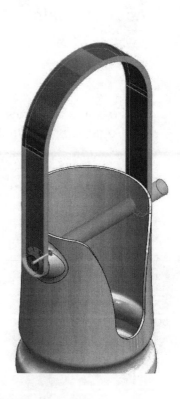

Click **OK.**

19. Adding a .030" fillet:

Click **Fillet**.

For Fillet Type, use the default **Constant Size**.

For Items to Fillet, select 4 edges of the handle.

For Radius, enter **.032in**.

Click **OK**.

Select 4 edges

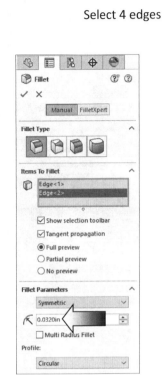

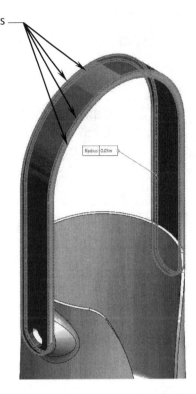

20. Creating a pin:

Open a **new sketch** on the Right plane.

Sketch the profile of the Pin as shown in the image.

Also add a vertical and a horizontal **centerline** to help define the sketch.

Add the dimensions and relations indicated to fully define the sketch.

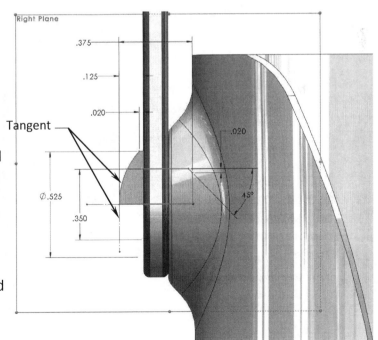

Switch to the **Features** tab.

Click **Revolved Boss-Base**.

Select the **Horizontal Centerline** as the Axis of Revolution.

For Angle, use the default **360°** angle.

Click **OK**.

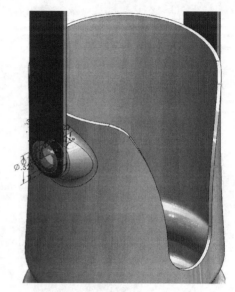

21. Mirroring the pin:

Click **Mirror** on the **Features** tab.

For Mirror Face/Plane, select the **Front** plane from the Feature tree.

Expand the **Bodies to Mirror** section and select the **Pin1** from the graphics area.

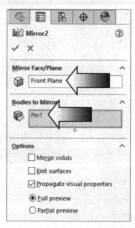

Click **OK**.

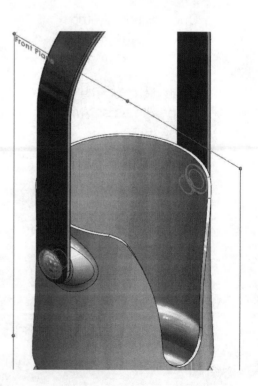

22. Creating an exploded view:

Select **Insert, Exploded View**.

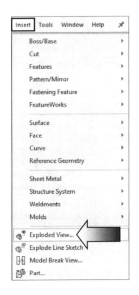

Select the **Pin1** either from the Feature tree or directly in the graphics area.

Drag the Z-Arrowhead outward approximately **1.500 inches**.

Click **Done**.

Repeat the same step to explode the other solid bodies. Click **OK** when done.

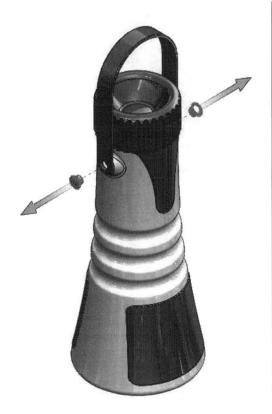

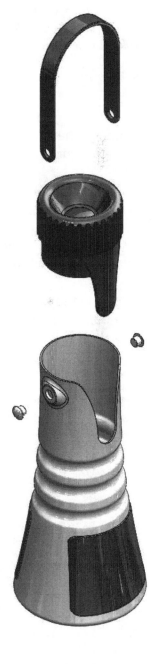

OPTIONAL:

The Exploded View must be <u>collapsed</u> in order to change the appearances of a multibody part.

(Switch to the Configuration-Manager tree, expand the Default and double-click on Exploded View1 to Collapse it).

A. Change the color of the **Handle** to **Black**.

B. Change the color of the **Cap** to **Black**.

C. Change the color of the **Bottle** to **White**.

D. Change the color of **2 Pins** to **White**.

E. Change the <u>face</u> color of the **2 labels** to **Black**.

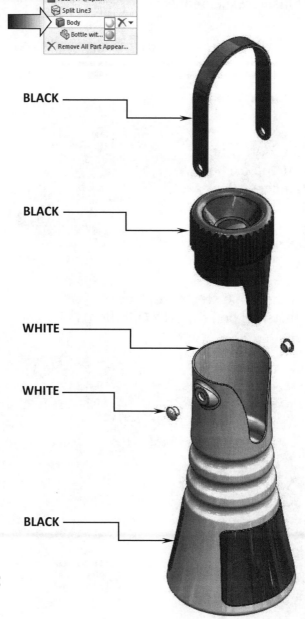

BLACK

BLACK

WHITE

WHITE

BLACK

23. Saving your work:

Select **Files, Save As**.

Enter: **Bottle_Multibody Design.sldprt** for the file name.

Click **Save**.

Close all documents.

Exercise: Coffee Mug – Multibody Design

The exercises give you an opportunity to apply what you have learned from the lessons. The instructions are limited to allow you to try things out using your own techniques, at your own pace.

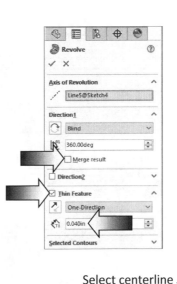

1. Opening a part document:

Browse to the Training Folder and open a part document named: **Mug_Multibody_Exe.sldprt**

This model contains 1 solid body and 4 sketches. One sketch will be used to create a solidbody and the other three will be used to create 3 surface bodies.

The surface bodies will be trimmed, knitted, and thickened into a solid body towards the completion of the design.

2. Creating a revolved <u>solid</u> body:

Switch to the **Features** tool tab.

Select **Sketch4** and click **Revolved Boss-Base.**

Click **NO** to leave the sketch open.

Select the **Centerline** in the middle of the sketch for Axis of Revolution.

<u>Clear</u> the **Merge Result** box.

Click **Thin Feature** and enter **.040in** for thickness.

Click **OK**.
This is the Insulation body.

Select centerline

3. Creating a revolved <u>surface</u> body:

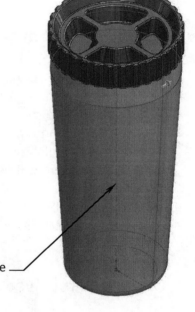

Switch to the **Surfaces** tool tab.

Revolved Surface

Revolves a sketch or sketch contours around an axis to create a surface feature. You can define the axis of revolution and the direction and extent of the revolved surface.

Select **Sketch5** and click **Revolved Surface**.

For Axis of Revolution, select the **Centerline** in the middle of the sketch.

Use the default **Blind** and **360°** revolve angle.

Expand the Feature Scope section and select: **Auto-Select**.

Select centerline

Click **OK**. This is the Mug body.

4. Creating the handle:

Click **Swept Surface**.

For Sweep Profile, select **Sketch7** from the Feature Manager tree.

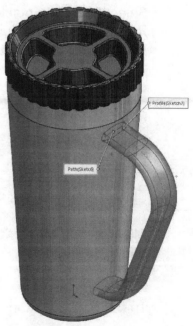

For Sweep Path, select **Sketch6**, also from the FeatureManager tree.

Click **OK**.
This is the Handle body.

5. Hiding the solid bodies:

For clarity, we will hide the solid bodies so that we can work with the surface bodies more easily.

Click the Solid <u>Bodies</u> folder and select **Hide**.

Both solid bodies, the Lid and the Insulation, are now hidden; only the two surface bodies are visible in the graphics area.

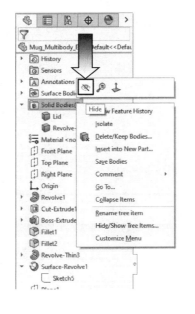

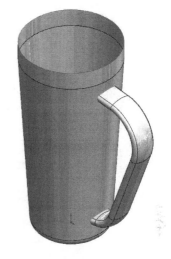

6. Trimming the surface bodies:

Click **Trim Surface**.

For Trim Type, select **Mutual Trim**.

For Selection, select both surface bodies, the **Mug** and the **Handle**.

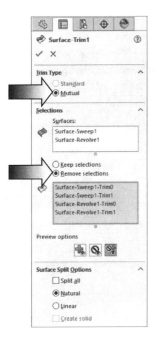

For Remove Selections, select the **inner portions** of the Handle and the **two elliptical surfaces** at the ends of the Handle as indicated.

The Trim automatically knits the two surfaces into a single surface body.

Click **OK**.

Remove 4 surfaces

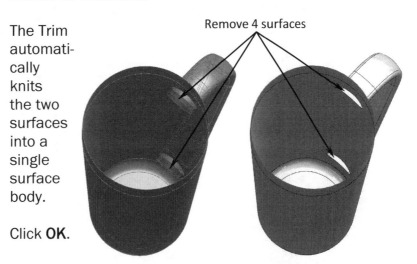

7. Adding Fillets:

Click **Fillet** on the **Surfaces** tab.

Use the default **Constant Size Fillet** option.

For Items To Fillet, select the **two edges** of the handle.

For Radius, enter: **.125in**.

Click **OK**.

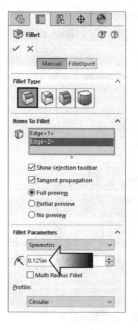

8. Thickening the surface body:

Click **Thicken** on the **Surfaces** tab.

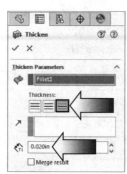

For Thicken Parameters, select the **surface body** in the graphics area.

For Direction, select the **Inside** button.

For Thickness, enter: **.020in**.

Click **OK**.

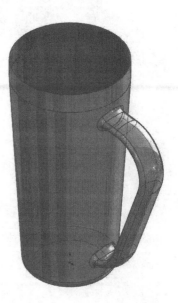

9. Renaming and changing color:

Expand the **Solid Bodies** folder and rename the solid bodies as following:

* **Lid**

* **Insulation**

* **Mug**

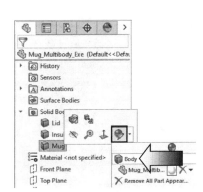

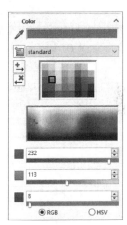

Click the name **Mug** and select: **Appearances, Body** (arrow).

In the Color Palette, select the **Brown** color, the 3rd color in the 2nd row.

Click **OK**.

Optional:
Change the color of the Insulation body to **White**.

10. Creating an exploded view:

Select **Insert, Exploded View**.

For Explode Step 1, select the **Mug** and drag the **Y-arrow** downward approximately **8.50in**.

Click **Done**.

For Explode Step 2, select the **Lid** and drag the **Y-arrow** upward, about **3.50in**.

Click **Done**.

Click **OK**.

Note: when the exploded view is active, all modifications will temporary be disabled. Simply collapse the exploded view to resume.

11. Saving your work:

Select **File, Save As**.

Enter: **Mug_Multibody_Exe Completed.sldprt** for the file name.

Click **Save**.
Close all documents.

Chapter 5: Surface Creation

Computer Mouse

Surface models are typically created by producing the surfaces one by one. These surfaces are trimmed to the final geometry then knitted and thickened into a solid model. The following lesson demonstrates these steps with a computer mouse model.

1. Creating the 1ˢᵗ sketch:

Select the Top plane and open a **new sketch**.

Sketch a **Centerline**, a **3-Point Arc,** then add the dimensions and relations shown.

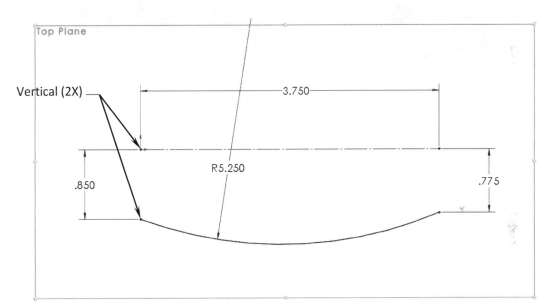

Change to the **Surfaces** tab and select the **Extruded Surface** command.

Use the default **Blind** type.

Enter **1.00in.** for Depth.

Click **OK**.

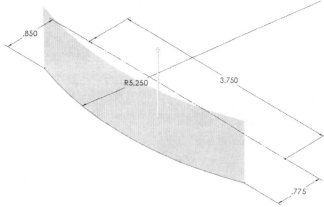

2. Creating the 2nd sketch:

Open a **new sketch** on the Front plane.

Sketch a **Centerline**, a **3-Point Arc,** then add the dimensions and relations shown below to fully define the sketch.

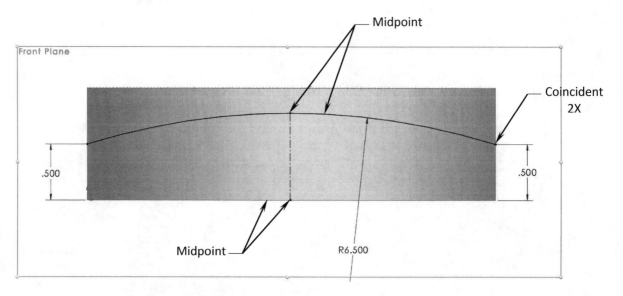

Switch back to the **Surfaces** tab and select the **Extruded Surface** command.

For Direction 1, use the default **Blind** type.

For Depth, enter **1.500in**.

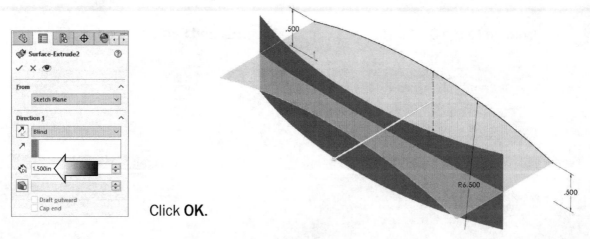

Click **OK.**

3. Trimming the surfaces:

The following references can be used when using the Standard Trim option: Surfaces, Sketch Entities, Curves, Planes, etc.

Select the **Trim Surface** command.

Click the **Standard Trim** option.

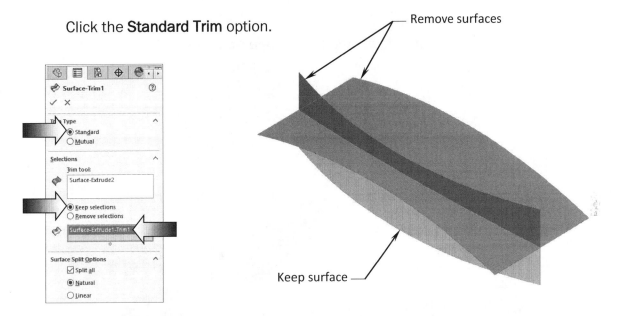

For Trim Tool, select the **Surface-Extrude2** from the graphics area.

For Keep Selections, select the lower portion of the **Surface-Extrude1**.

Click **OK**.

Hide the upper surface as indicated.

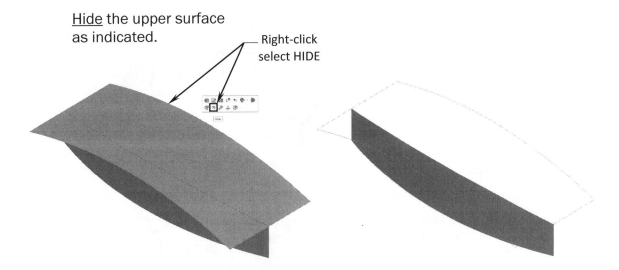

4. Mirroring a surface body:

Change to the **Features** tab.

Mirroring Bodies:
Surfaces are recognized as bodies.
They should be mirrored as bodies, not as features.

Select the **Mirror** command.

For Mirror Face/Plane, select the **Front** plane from the FeatureManager tree.

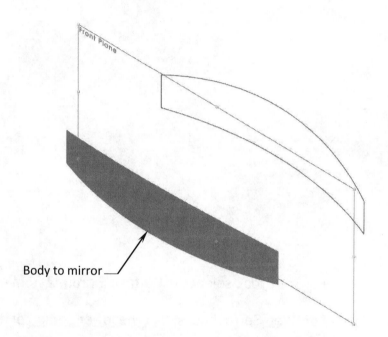

Body to mirror

Expand the **Bodies to Mirror** section and select the **Surface-Trim1** directly from the graphics area.

Clear the **Merge Solid** and **Knit Surfaces** checkboxes.

Click **OK**.

There should be two surfaces listed inside the Surface Bodies folder on the Feature tree.

5. Making the 1ˢᵗ loft profile:

A loft feature requires two or more profiles and one or more guide curves.

Open a <u>new</u> **3D-Sketch** and sketch a **3-Point Arc** as shown below.

Ensure that both ends of the arc are **Coincident** to the vertices of the two surfaces.

Add the dimensions / relations shown.

<u>Exit</u> the 3D Sketch (or press Control+Q).

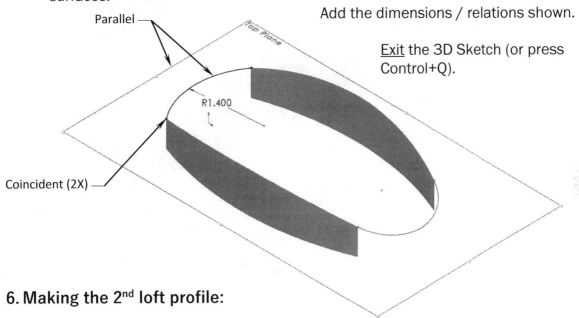

6. Making the 2ⁿᵈ loft profile:

Open another **3D-Sketch** and sketch the second **3-Point Arc** on the right end of the model.

Ensure both ends of the arc are **Coincident** to the vertices of the two surfaces.

Add a **Tangent** relation between the arc and the upper edge of the side surface as noted.

<u>Exit</u> the 3D Sketch (Control+Q).

7. Making the 3rd loft profile:

Start another **3D-Sketch** once again.

Sketch a **3-Point Arc** and add two **Coincident** relations to both ends.

Right Plane

R6.000

1.875

Coincident (2X)

Add the radius dimension to fully define the sketch.

Exit the 3D-Sketch.

8. Creating a lofted surface:

Helpful Tips

Lofted Surface:
Loft creates a feature by making transitions between profiles. Two profiles or more are required. Only the first, last, or first and last profiles can be points. The rest can be any shape or size.

Select the **Lofted Surface** command.

For Loft Profiles, select the **3 arcs** as indicated.

For Guide Curves, select the **2 Upper** edges of the two surfaces.

Keep all other parameters at their defaults.

Click **OK**.

3 Loft profiles

2 Guide Curves

9. Making the 1st fill profile:

Additional geometry must be created to form the boundaries needed to close off the two ends.

Open a **new sketch** on the Top plane.

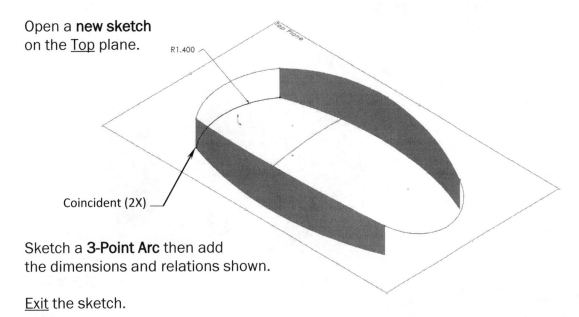

R1.400

Coincident (2X)

Sketch a **3-Point Arc** then add the dimensions and relations shown.

Exit the sketch.

10. Making the 2nd fill profile:

Open another **sketch** on the Top plane.

Sketch a **3-Point Arc** on the right end.

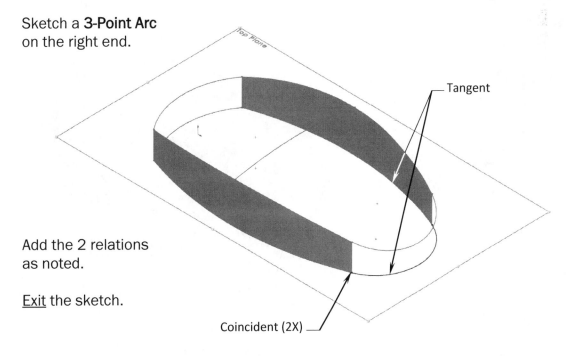

Tangent

Add the 2 relations as noted.

Exit the sketch.

Coincident (2X)

11. Creating 2 filled surfaces:

Select the **Fill Surface** command.

Select the **4 edges** on the right end of the surface model.

> **Filled Surface:**
> The Filled Surface command constructs a surface patch with any number of sides, within a boundary defined by existing model edges, sketches, or curves, including composite curves.

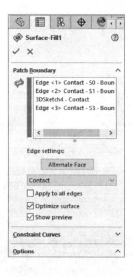

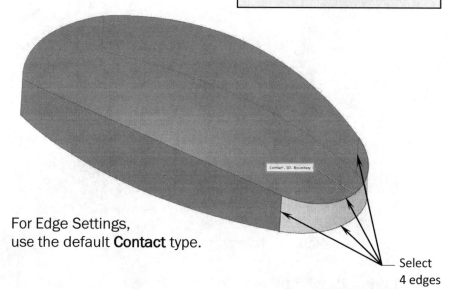

For Edge Settings, use the default **Contact** type.

Select 4 edges

Click **OK**.

Rotate the surface model and fill its left side.

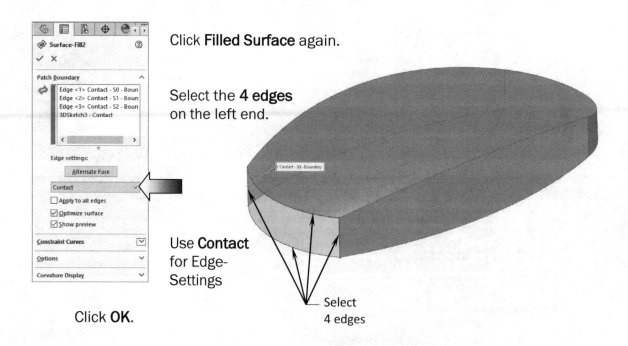

Click **Filled Surface** again.

Select the **4 edges** on the left end.

Use **Contact** for Edge-Settings

Select 4 edges

Click **OK**.

12. Creating a planar surface:

The bottom of the surface model is still open. It can easily be patched with the Planar Surface tool.

Click **Planar Surface.**

Select 4 bottom edges

Select **4 edges** at bottom of the surface model.

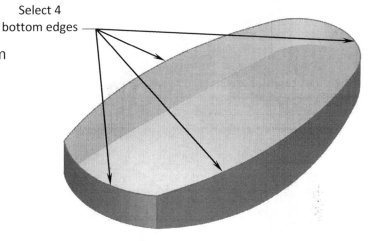

Click **OK.**

13. Knitting the surfaces:

The Knit Surface command combines two or more surfaces into one.

Gap Control:
Displays the edge pairs that might introduce gap problems. The gap range depends on the knitting tolerance. Only gaps within the selected gap range are listed. You can modify the gap range if required.

Helpful Tips

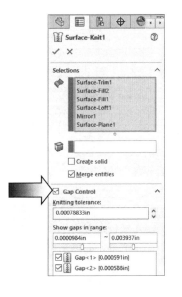

Click **Knit Surface.**

Select all **6 surfaces** in the graphics area.

Enable the **Gap Control** option.

Click **OK.**

14. Creating the 1.00in. fillets:

Rotate the surface model to its left side.

Click **Fillet**.

Use the default **Constant Size** option.

Enter **1.000in** for radius size.

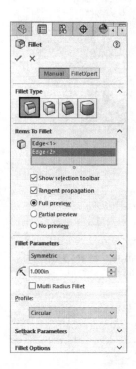

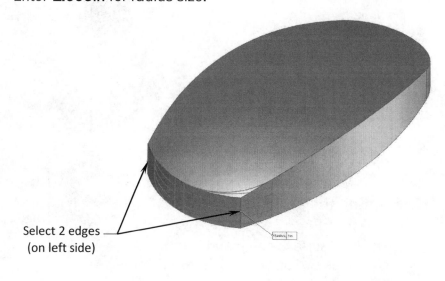

Select 2 edges
(on left side)

Select the **2 vertical edges** as indicated.

The preview graphics shows
2 fillets are being created.

Keep all other parameters
at their defaults.

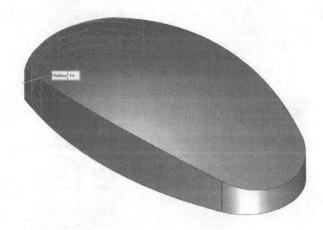

Click **OK**.

15. Creating a split line feature:

Open a **new sketch** on the <u>Front</u> plane.

Sketch a **2-point Spline** from right to left. Only the right end is locked at the midpoint of the surface model, not the left end.

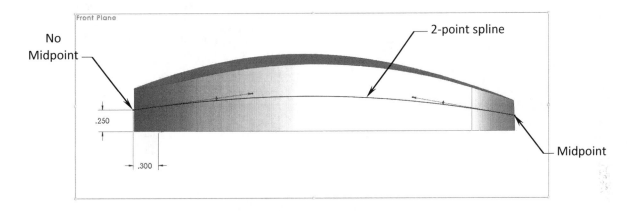

Add the dimensions / relations shown to fully define the sketch.

Select the **Split Line** command under the **Surfaces, Curves** drop-down menu.

Helpful Tips

Split Lines:
The Split Line tool projects a sketch to surfaces and divides a selected face into multiple separate faces.

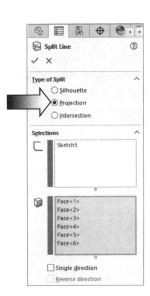

For Type of Split, select **Projection**.

For Selections, the **Spline** should be selected already.

For Faces to Split, select the **6 faces** as noted.

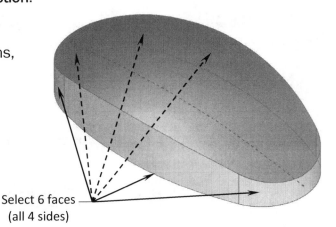

Select 6 faces (all 4 sides)

Click **OK**.

16. Creating a face fillet with hold line:

When using the Face Fillet with Hold Line, the radius of the fillet is driven by the distance between the hold line and the edge to fillet.

Click **Fillet**.

> **Hold Line:**
> Creates a face fillet whose shape is determined by a part edge or projected split line.

For Fillet Type, select the **Face Fillet** option.

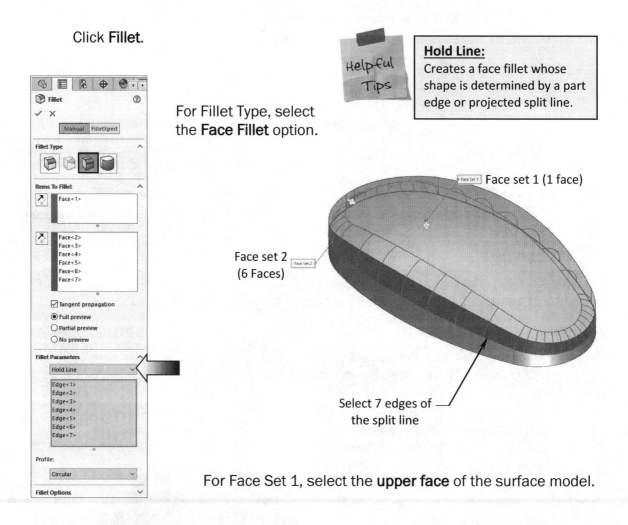

Face set 1 (1 face)

Face set 2 (6 Faces)

Select 7 edges of the split line

For Face Set 1, select the **upper face** of the surface model.

For Face Set 2, select **7 faces** above the split line.

For Fillet Parameters, select the **Hold Line** option.

For Hold Line Edges, select the **7 edges** of the split line.

Click **OK**.

17. Creating the .030" fillets:

Click **Fillet** once again.

Select the **Constant Size Radius** option.

Enter **.030in** for radius size.

Select **1 edge** at the bottom. The default Tangent-Propagation allows the fillet to run along the perimeter of the bottom surface.

Select edge

Click **OK**.

18. Saving your work:

Select **File, Save As**.

Enter **Computer Mouse_Completed.sldprt** for the file name.

Click **Save**.

OPTIONAL:
Change the upper faces to White color.
Change the lower faces to Dark Gray color and enable the Realview Graphics option.

Boundary & Lofted Surface Exercise
Phone Case

1. Opening a part document:

Open a part document named:
Phone Case.sldprt

This exercise was developed to show the flexibilities of the Lofted Surface over the Boundary Surface command.

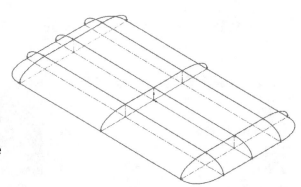

2. Creating a boundary surface:

The boundary surface feature often produces a higher quality surface than the lofted surface feature, but it does not have the same flexibilities as the loft.

Click **Boundary Surface**.

For Direction 1, select the **2 sketch profiles** as labeled.

For Direction 2, select the **3 sketch profiles** as noted below.

Notice the Surface Boundary is using the entire length of the 3 profiles to fill the boundary.

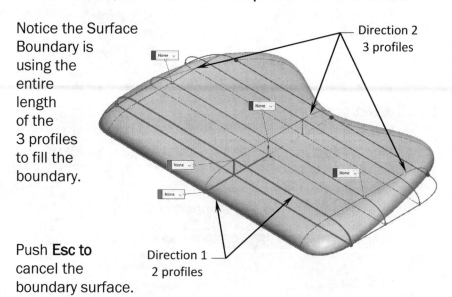

Direction 2
3 profiles

Direction 1
2 profiles

Push **Esc to** cancel the boundary surface.

3. Creating the 1st lofted surface:

We will now try out the lofted surface command.

Click **Lofted Surface**.

For Loft Profiles, select the **2 sketch profiles** as labeled.

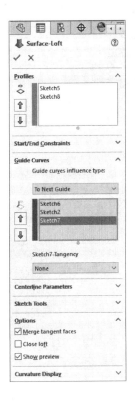

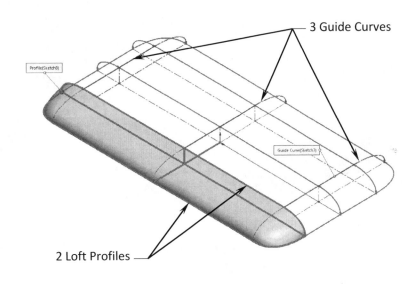

3 Guide Curves

2 Loft Profiles

For Guide Curves, select the **3 sketch profiles** as noted above.

Notice the loft only uses
a part of the Guide
Curves to close off
the boundary, not
the entire length.

Keep all other parameters
at their default values.

Click **OK**.

4. Creating the 2nd lofted surface:

It is easiest to mirror the lofted surface. However, for practice purposes, we will once again recreate the lofted surface for the opposite side.

Click **Lofted Surface**.

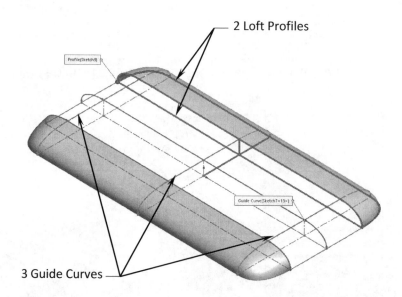

2 Loft Profiles

3 Guide Curves

For Loft Profiles, select the **2 sketch profiles** as indicated above.

For Guide Curves, select the **3 sketch profiles** as labeled.

Keep all other parameters at their default values.

Click **OK**.

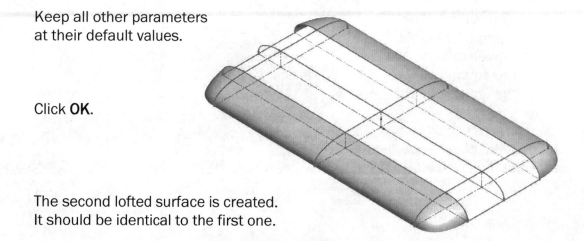

The second lofted surface is created. It should be identical to the first one.

5. Creating the 3rd lofted surface:

The center section can be filled using several different options. For the purpose of this exercise, we will continue to use the lofted surface command.

Click **Lofted Surface**.

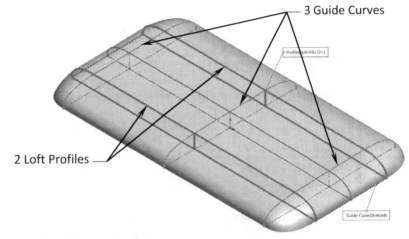

3 Guide Curves

2 Loft Profiles

For Loft Profiles, select the **2 sketch profiles** as noted above.

For Guide Curves, select the **3 sketch profiles** as labeled in the image above.

Note: The center sketch profile can also be used as one of the loft profiles; however, it does not make any significant changes. Therefore, we will leave it as is.

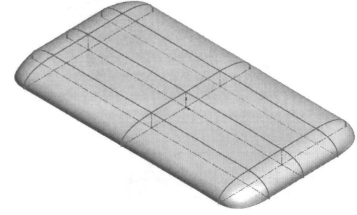

Again, leave all other parameters at their default values.

Click **OK**.

6. Hiding the sketches:

After the surface body is created, we no longer need to see the sketches.

Hold the **Control** key and select <u>all sketches</u> from either the FeatureManager tree or in the graphics area.

Release the Control key and click **Hide** (arrow).

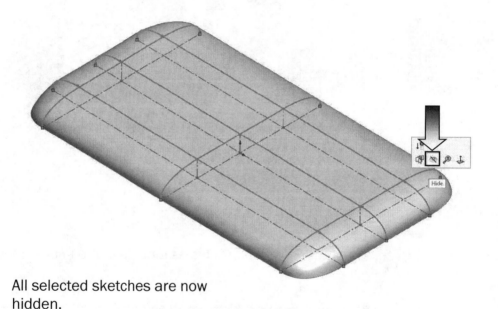

All selected sketches are now hidden.

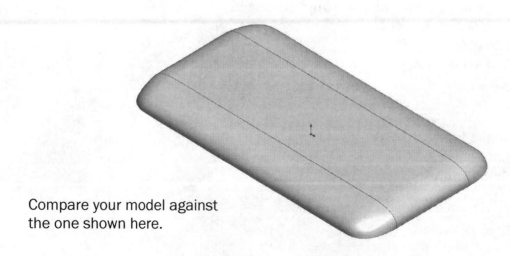

Compare your model against the one shown here.

7. Knitting the surface:

The last step is to knit all surfaces into a single surface.

Click **Knit Surface**.

For Selections, select all **3 surfaces** in the graphics area.

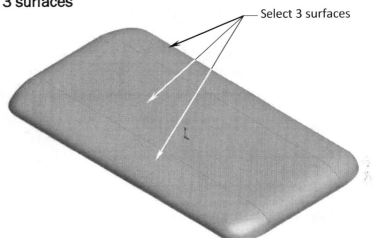

Select 3 surfaces

Enable the **Merge Entities** checkbox (arrow).

Click **OK**.

8. Saving your work:

Select **File, Save As**.

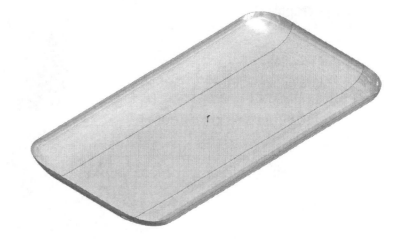

Enter **Phone Case_ Completed.sldprt** for the file name.

Click **Save**.

Exercise: Trim & Mirror Surfaces

The exercises give you an opportunity to apply what you have learned from the lessons. The instructions are limited to allow you to try things out using your own techniques, at your own pace.

1. Opening a part document:

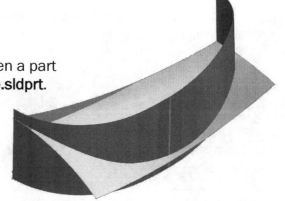

Browse to the Training Folder and open a part document named: **Lesson 4_Exercise.sldprt**.

There are 2 surfaces in this model. They will be used to trim the geometry to its final shape and size.

2. Trimming the surfaces:

Click **Trim Surface**.

For Trim Type, select **Mutual**.

For Trimming Surfaces, select the **2 surfaces** in the graphics area.

For Remove Selections, click the 2 outer portions of the 2 surfaces as indicated.

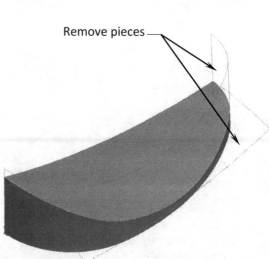

Remove pieces

For Surface Split Option, select **Natural**.

Click **OK**.

3. Mirroring the surface body:

Switch to the **Features** tab.

Click **Mirror**.

For Mirror Face/Plane select the **Right** plane from the Feature tree.

For Bodies to Mirror, select the **surface body** in the graphics area as noted.

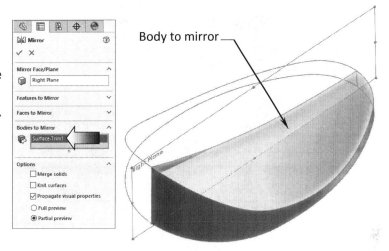

Click **OK**.

4. Adding a variable size fillet:

Click **Fillet**.

For Fillet Type, select **Variable Size Fillet**.

Select the <u>upper edge</u> of the surface model as noted.

Enter the fillet values of **.375in** on the left side, and **.080in** on the right side.

Click **OK**.

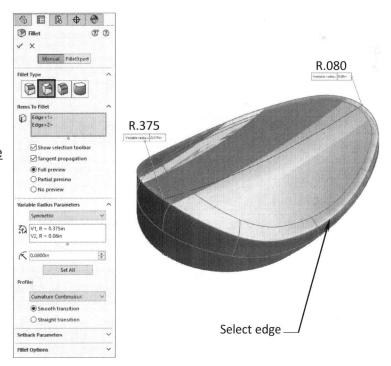

5. Saving your work:

Select **File, Save As**.

Enter **Lesson 5_Exercise_Completed** for the file name.

Click **Save**.

Close all documents.

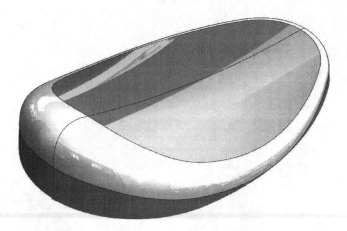

Chapter 6: Using Filled, Knit & Boundary Surface

Catheter Handle

1. Opening a part document:

Open a part document named:
Boundary Surface.sldprt

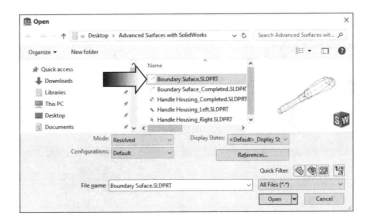

This lesson will teach us the use of the Boundary Surface command and other frequently used surfacing tools such as: Surface Revolved, Filled, Split, and Knit.

This part document has several sketches that have already been created to help focus on the use of the surfacing tools. They are labeled as Direction 1 and Direction 2.

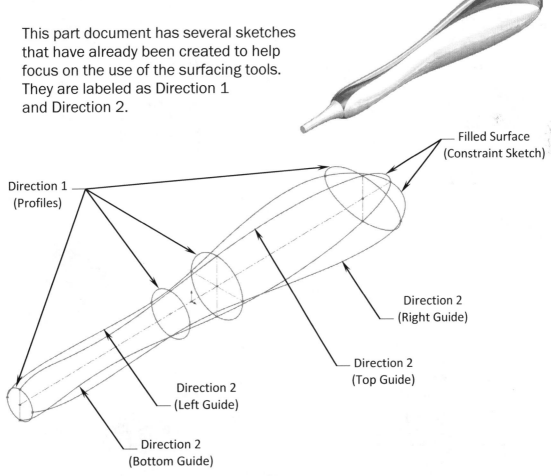

Filled Surface
(Constraint Sketch)

Direction 1
(Profiles)

Direction 2
(Right Guide)

Direction 2
(Top Guide)

Direction 2
(Left Guide)

Direction 2
(Bottom Guide)

2. Creating a boundary surface:

Click **Boundary Surface** on the **Surfaces** tab.

For Direction 1, select the **4 elliptical sketches** from the graphics area.

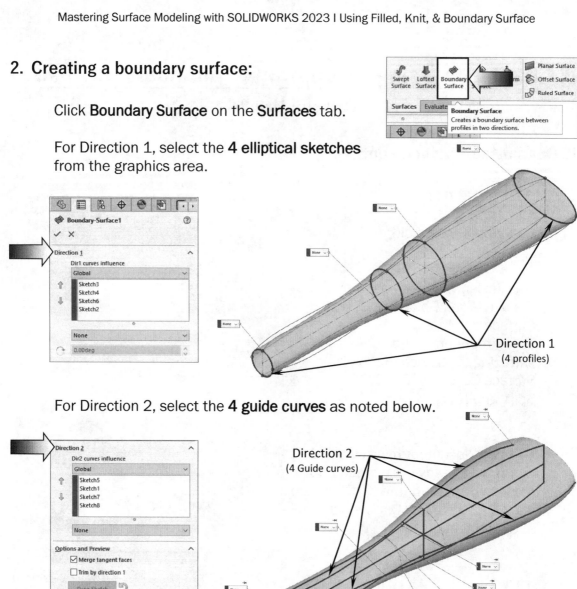

Direction 1
(4 profiles)

For Direction 2, select the **4 guide curves** as noted below.

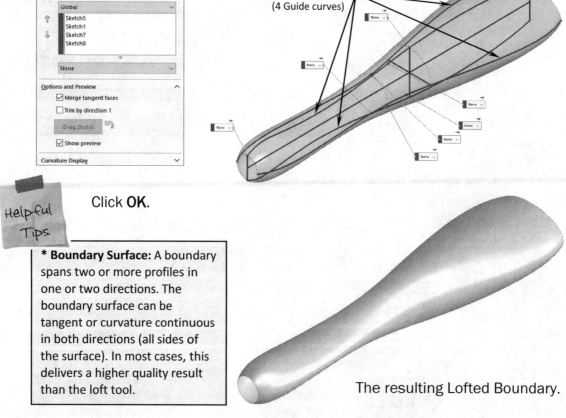

Direction 2
(4 Guide curves)

Click **OK**.

Helpful Tips

* Boundary Surface: A boundary spans two or more profiles in one or two directions. The boundary surface can be tangent or curvature continuous in both directions (all sides of the surface). In most cases, this delivers a higher quality result than the loft tool.

The resulting Lofted Boundary.

3. Creating a revolved surface:

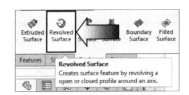

Select **Sketch9** on the FeatureManager tree and click **Edit Sketch**.

The sketch has a vertical line and a 2-point Spline. It has already been fully defined.

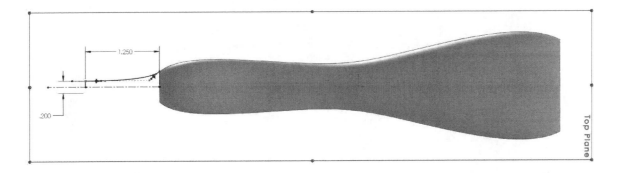

Switch to the **Surfaces** tab and click **Revolved Surface**.

For the Axis of Revolution, select the **Centerline** as indicated.

Use the default **Blind** type and **360°**.

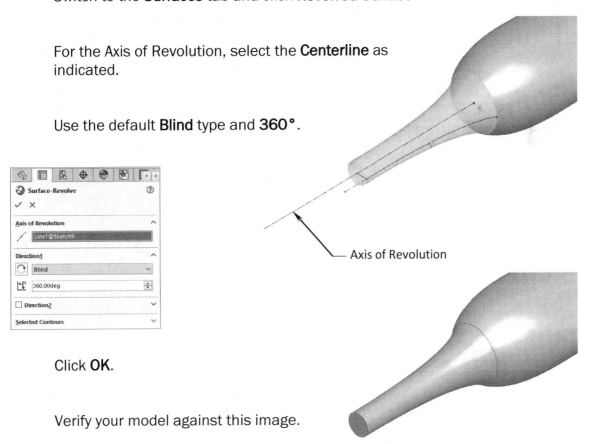

Axis of Revolution

Click **OK**.

Verify your model against this image.

4. Creating a filled surface:

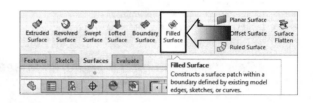

The right end of the surface model will be filled with two constraint sketches.

Click **Filled Surface**.

For Patch Boundary, select the **elliptical edge**.

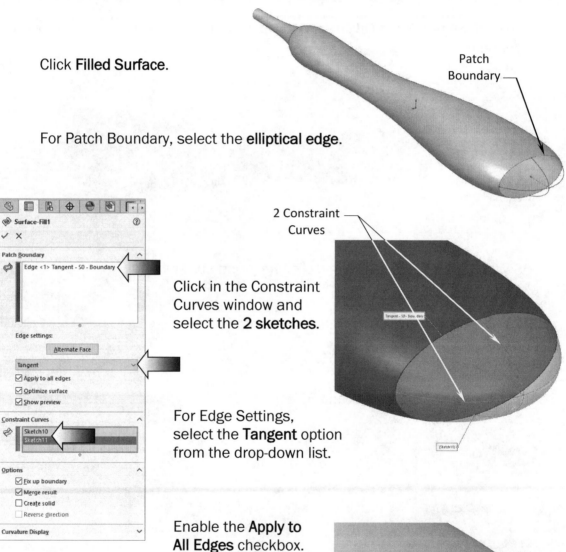

Patch Boundary

2 Constraint Curves

Click in the Constraint Curves window and select the **2 sketches**.

For Edge Settings, select the **Tangent** option from the drop-down list.

Enable the **Apply to All Edges** checkbox.

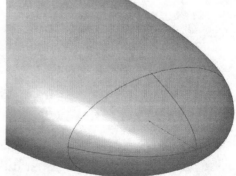

Also enable the **Fix Up Boundary** and the **Merge Result** checkboxes.

Click **OK**.

Click the constraint sketch
and select **HIDE** (arrow).

Hide both constraint sketches.

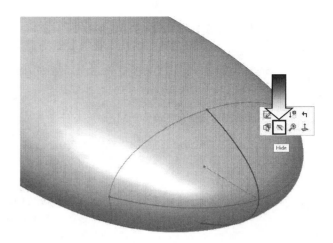

5. Knitting the surfaces:

Sometimes it is faster and easier to
create features as solid features rather
than surface features.

Because of this, the surface model will now
be knitted and converted into a solid model.

Click **Knit Surface**.

Select **all surfaces** from the
graphics area.

Enable the **Create Solid** and
Merge Entities checkboxes.

Click **OK**.

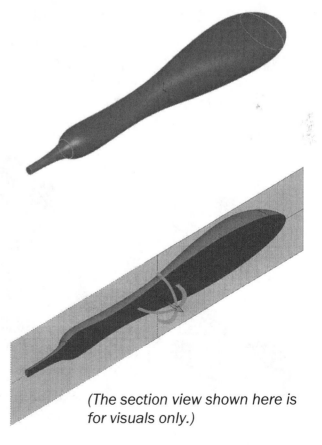

*(The section view shown here is
for visuals only.)*

6. Creating a split line feature:

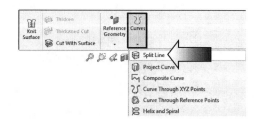

Select the **Split Line** command from the **Surfaces, Curves** drop-down menu.

For Type of Split, select **Projection**.

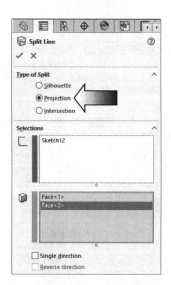

For Selections, select **Sketch12** from the FeatureManager tree.

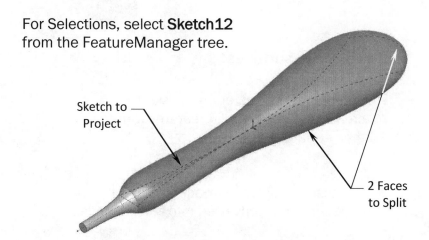

Sketch to Project

2 Faces to Split

For Faces to Split, select the **Boundary Surface** and the **Filled Surface** as noted.

Click **OK**.

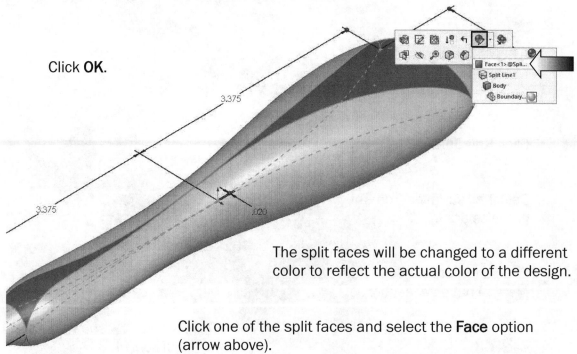

The split faces will be changed to a different color to reflect the actual color of the design.

Click one of the split faces and select the **Face** option (arrow above).

7. Changing the face color:

Using the Color Palette, select the (Navy) **Blue** color.

Change the **Red** value to **0**.

Set the **Green** value to **125**.

Set the Blue value to **255**.

Click **OK**.

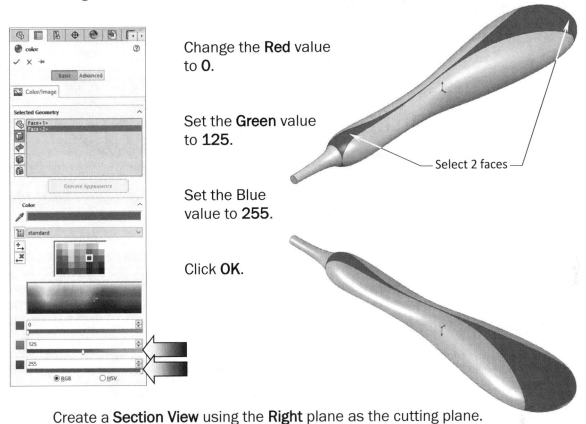

Select 2 faces

Create a **Section View** using the **Right** plane as the cutting plane.

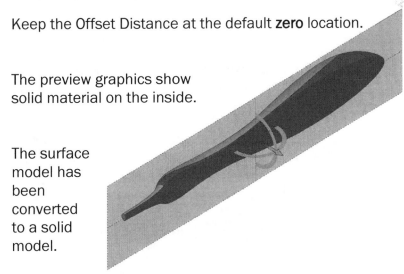

Keep the Offset Distance at the default **zero** location.

The preview graphics show solid material on the inside.

The surface model has been converted to a solid model.

(Keep the section view on for the next step.)

8. Shelling the model:

Switch to the **Features** tab and click the **Shell** command.

For Wall Thickness, enter **.060in**.

For Face to Remove, select the **face** at the left end of the model.

Face to remove

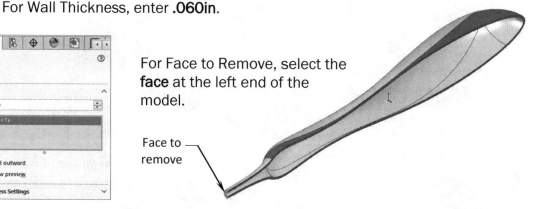

Click **OK** and click-off the section view command.

9. Enabling RealView Graphics:

RealView Graphics is hardware (graphics card) support of advanced shading in real time, including self-shadowing and scene reflections.

When applying an appearance, use RealView Graphics for a realistic model display. RealView Graphics is available with supported graphics cards only.

Enable the **RealView Graphics** option under **View Settings** drop-down menu.

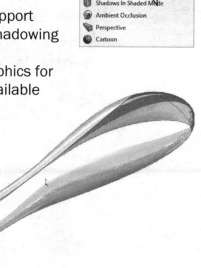

10. Saving your work:

Select **File, Save As**.

Enter **Boundary Surface_Completed.sldprt** for the file name.

Click **Save**.

Using Deform Surface

1. Opening a part document:

Open a part document named:
Surface Deformed_Exe.sldprt

This part document has four visible sketches and two hidden sketches. The visible sketches will be used to create the main surface body and the hidden sketches will be used to trim and deform the main surface body to its final size and shape.

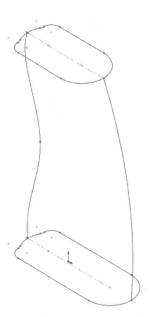

2. Creating a lofted surface:

Switch to the **Surfaces** tab and click: **Lofted Surface**.

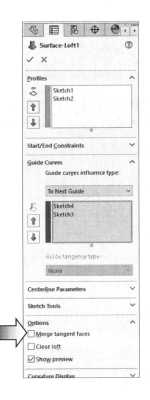

For Loft Profiles, select **Sketch1** and **Sketch2** (top and bottom) from the graphics area.

Expand the Guide Curve section and select: **Sketch3** and **Sketch4** (left and right) also from the graphics area.

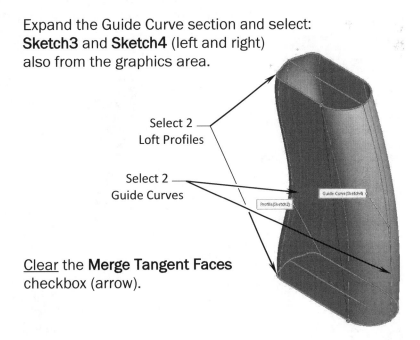

Select 2 Loft Profiles

Select 2 Guide Curves

Clear the **Merge Tangent Faces** checkbox (arrow).

Click **OK**.

3. Creating a deform feature:

Aside from using the deform feature with a Point or a body, the Curve to Curve deform function is a precise method for altering complex shapes. We can deform the surface body by mapping geometry from initial curves to a set of target curves.

Select **Insert, Features, Deform** (arrow).

Click the **Curve to Curve** option.

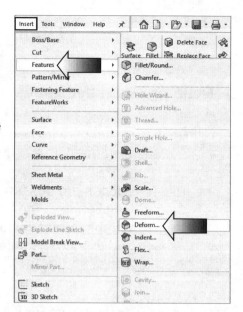

For Initial Curves, select the **outer edge** on the <u>right side</u> of the surface body as noted.

For Target Curves, select the **Finger Grips** sketch from the FeatureManager tree.

For Fixed Faces, select the **2 faces** on the <u>left side</u> as indicated.

Drag the **Shape Accuracy** all the way to the right to increase the surface quality (this setting may decrease performance).

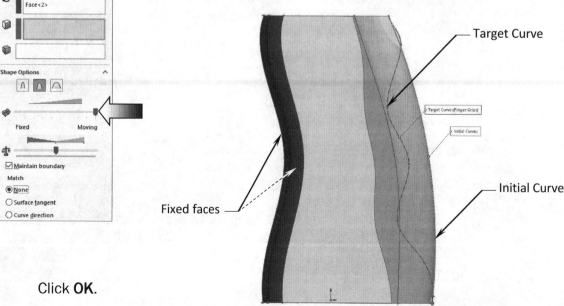

Target Curve

Initial Curve

Fixed faces

Click **OK**.

4. Trimming to the final size:

Click **Trim Surface**.

For Trim Type, select **Standard**.

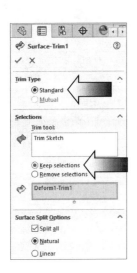

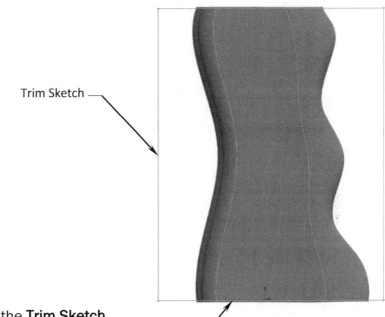

Trim Sketch

Top & bottom are not flat

For Trim Tool, select the **Trim Sketch** from the FeatureManager tree.

Click the Keep Selections option and select the **surface body** <u>inside</u> the rectangular sketch.

Keep all other parameters at their default settings.

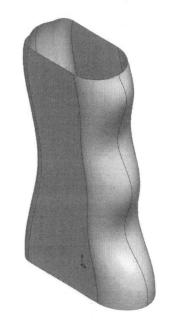

Click **OK**.

Compare your surface model against the image shown on the right.

5. Patching the bottom surface:

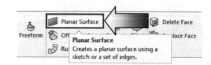

Click **Planar Surface**.

For Bounding Entities, select all **6 edges** at the bottom of the surface body.

The preview graphics show a new surface is being created to close-off the opening at the bottom.

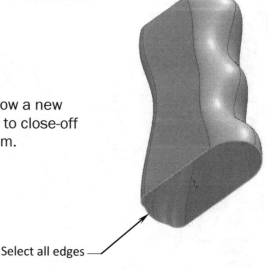

Click **OK**.

Select all edges —

6. Knitting the surfaces:

Click **Knit Surface**.

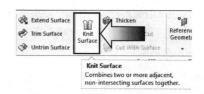

Select **both surfaces** from the graphics area.

Enable the **Merge Entities** checkbox.

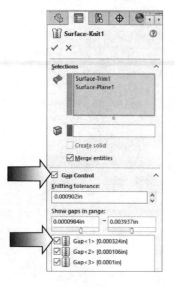

Select the **Gap Control** checkbox and click one of the **Gap** checkboxes to allow small gaps to be healed automatically.

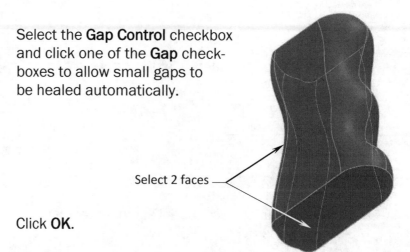

Select 2 faces —

Click **OK**.

7. Adding fillets:

Click **Fillet**.

Use the default **Constant Size** radius option.

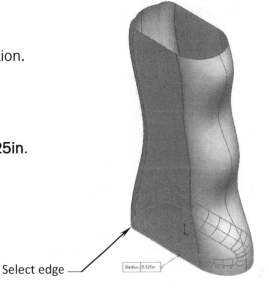

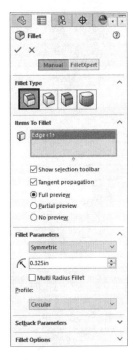

For radius size, enter **.325in**.

Select edge ⎯

Radius: 0.325in

For Items to Fillet, select **one of the edges** at the bottom of the surface model as noted.

Click **OK**.

8. Saving your work:

Select **File, Save As**.

Enter: Surface **Deformed_Completed.sldprt** for the file name.

Click **Save**.

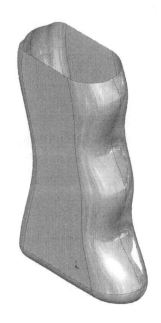

OPTIONAL: Drag the Rollback line <u>below</u> the **Surface-Plane**1 and add the following:

9. Adding a planar surface:

Add a planar surface on the top of the surface model by selecting all <u>6 edges</u> on top.

10. Knitting the surfaces:

Move the Rollback line <u>below</u> the **Surface-Knit1** and <u>edit</u> the **Surface-Knit1**.

Enable the **Create Solid** checkbox.

Select **all surfaces** and enable the **Gap** checkboxes.

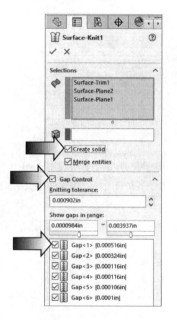

11. Shelling the model:

Drag the Roll-back line <u>below</u> the **Fillet1**.

Add a **Shell** feature with a thickness of **.070in**.

Resave the model.

Exercise: Boundary Surface

The exercises give you an opportunity to apply what you have learned from the lessons. The instructions are limited to allow you to try things out using your own techniques, at your own pace.

1. Opening a part document:

Browse to the Training Folder and open a part document named: **Lesson 5_Exercise.sldprt**.

There are 3 Sketch Profiles, 3 Guide Curves, and 2 additional Boundary Profiles in this model.

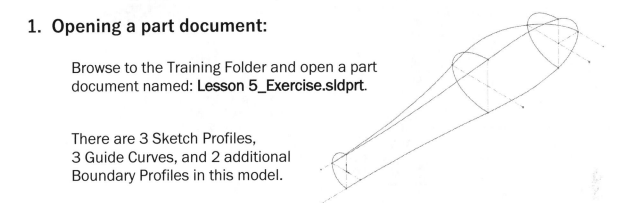

The Boundary Surface tool creates surfaces that can be tangent or curvature continuous in both directions (all sides of the surface). This tool can deliver a higher quality result than the loft tool.

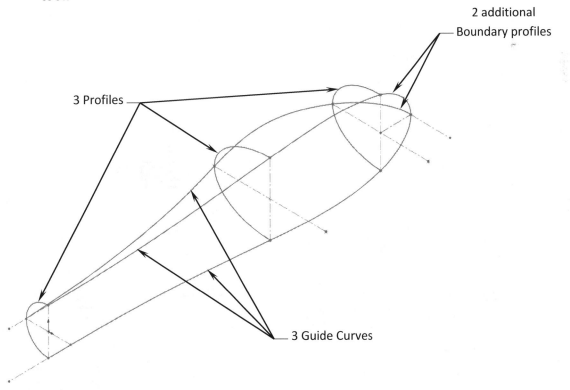

2 additional
Boundary profiles

3 Profiles

3 Guide Curves

2. Creating the 1ˢᵗ Boundary Surface:

Click **Boundary Surface**.

For Direction 1
select the **3
Sketch Profiles**
either from the
Feature tree
or from the
graphics area.

For Direction 2
select the **3
Guide Curves**
from the
graphics area.

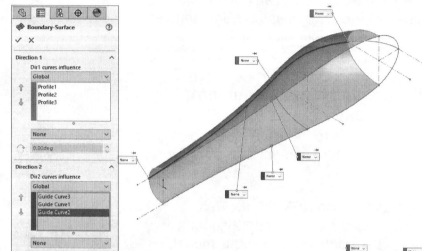

Click **OK**.

3. Creating the 2ⁿᵈ Boundary Surface:

Click **Boundary Surface** once again.

For Direction 1
select the **edge**
on the right
end of the
surface model
and the sketch
Profile4.

For Direction 2
select the sketch
Profile 5.

Select **Edge1** in
the Direction 1
section and
change the
Tangent Type to
Tangency to Face.

Click **OK**.

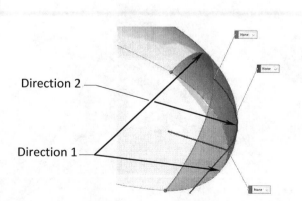

Direction 2

Direction 1

4. Mirroring the surface bodies:

Switch to the **Features** tab.

Click **Mirror**.

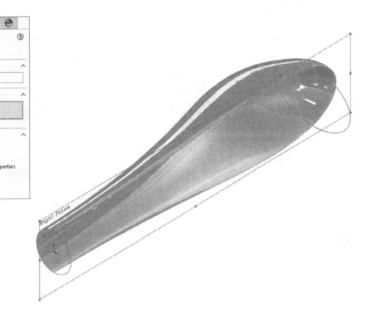

For Mirror Face select the **Right** plane.

For Bodies to Mirror, select **both Surface Bodies** from the graphics area.

Enable the **Knit Surfaces** checkbox.

Click **OK**.

5. Saving your work:

Select **File, Save As**.

Enter **Lesson 5_Exercise_Completed** for the file name.

Click **Save**.

Close all documents.

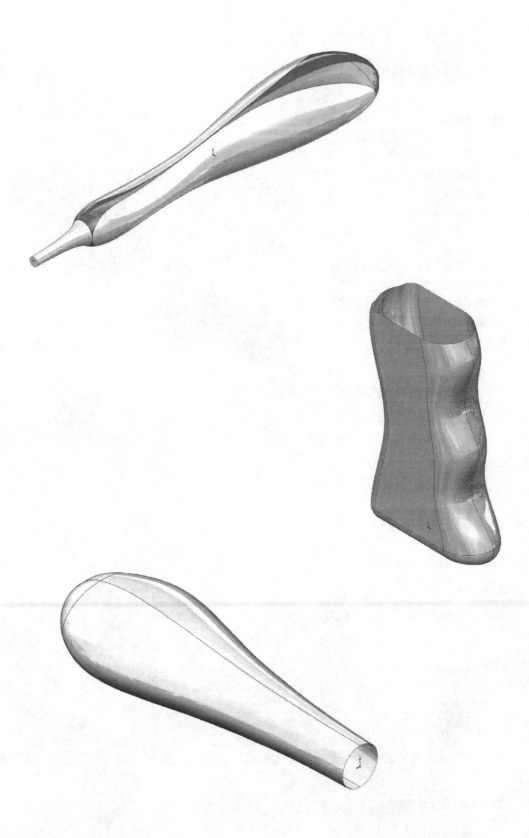

Mid-Term Quiz

Toy Design

Instructions will be limited so that you can apply what you have learned from the previous lessons to create this surface model.

1. Opening a part document:

Browse to the Training Folder and open a part document named: **Mid-Term Quiz.sldprt**

There are 2 surface bodies and 2 sketches that have been previously created.

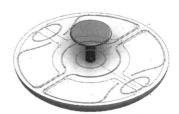

You will need to perform the following tasks:

1. Use **Sketch3** and **Sketch4** and make 2 different Split features.

2. Create 2 recess features using the 2 Ellipses in **Sketch3**.

3. Create a mirror-body of the Lower Surface.

4. Calculate the **Total Surface Area** of the two surface bodies

Notes:

Only follow the instructions in this lesson when needed. You can use any methods that you want to create the surfaces as long as they have the same size and location dimensions; and most importantly, they must have the correct Surface Areas as the answers to the final questions.

2. Creating the 1st trimmed surface:

Change to the bottom view orientation (Contro+6). <u>Hide</u> the **Center Surface**.

Switch to the **Surfaces** tab.

Hide the
Center surface

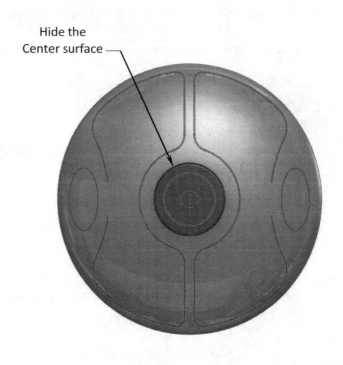

Click **Trim Surface**.

Select 5 regions

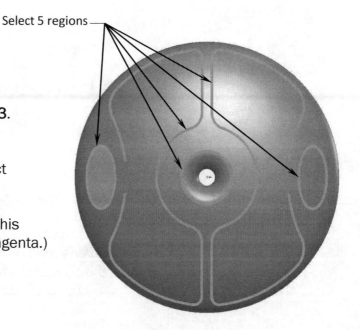

For Trim Type, select **Standard**.

For Trim Tool, select **Sketch3**.

For Remove Selection, select the **5 regions** as indicated. (Only select the trim regions when they change color; in this case, the default color is Magenta.)

Click **OK**.

3. Creating the 2nd trimmed surface:

Change to the **Top** view orientation (Control+5).

Click **Trim Surface**.

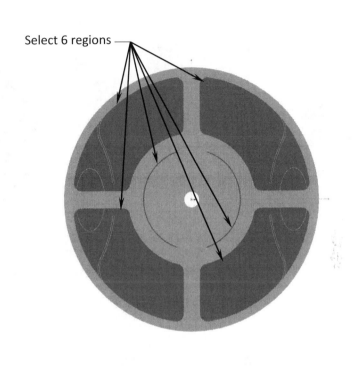

Select 6 regions

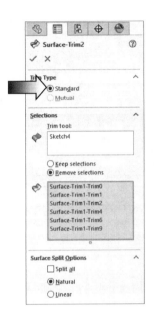

For Trim Type, click the **Standard** option.

For Trim Tool, select **Sketch4**.

For Remove Selections, select the **6 regions** as noted above.

Click **OK**.

Inspect your surface model against the image shown on the right. The Total Surface Area will not be right if one or more regions are missed.

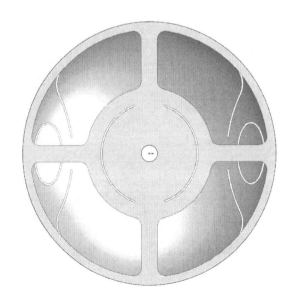

4. Making an offset surface:

The offset surfaces will be used to create the recess features in step 6 and step 7.

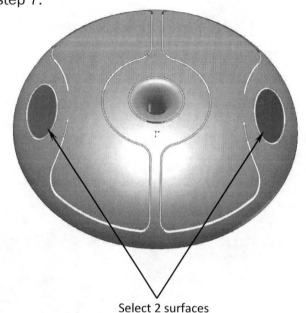

Click **Offset Surface.**

For Offset Parameters, select the **2 elliptical surfaces** as indicated.

Select 2 surfaces

For Offset Distance, enter **.050in**.

The 2 offset surfaces must be placed <u>inside</u> of the surface model.

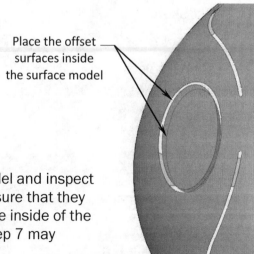

Place the offset surfaces inside the surface model

Click **OK**.

Rotate the surface model and inspect the offset surfaces. Ensure that they are intact, placed on the inside of the model or step 6 and step 7 may fail.

5. Deleting the reference surfaces:

The initial elliptical surfaces are no longer needed; they can be deleted.

Expand the **Surface Bodies** folder.

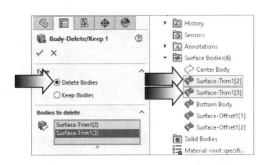

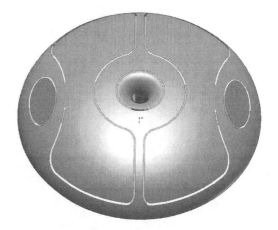

Right-click the **Surface-Trim1** and select: **Delete/Keep Body**.
Also select the 2nd **Surface-Trim** from the Surface Body folder to delete.

For Type, click **Delete Bodies**.

Click **OK**.

6. Creating the 1st lofted surface:

The recess feature can now be created by connecting the two elliptical edges.

Click **Lofted Surface**.

For Loft Profiles, select the **elliptical edges** as indicated.

The blue-connectors should be right next to each other to prevent twisting. Drag the connectors to reposition them, if needed.
Click **OK**.

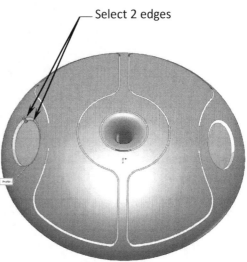

Select 2 edges

7. Creating the 2ⁿᵈ lofted surface:

Click **Lofted Surface** again.

For Loft Profiles, select the **two elliptical edges** on the other side.

Drag the blue connectors if needed, to keep the loft feature from twisting.

Click **OK**.

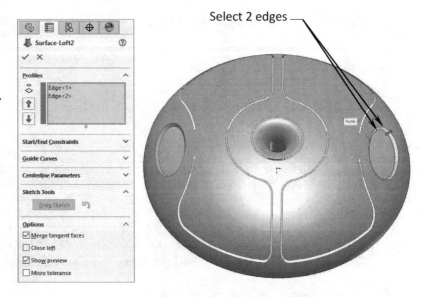

Select 2 edges

8. Knitting the surfaces:

In order to add fillets between the surfaces, they need to be knitted into a single surface.

Click **Knit Surface**.

For Selections, select **all** surfaces <u>except</u> for the Center Surface.

Enable the **Gap Control** checkbox.

Click **OK**.

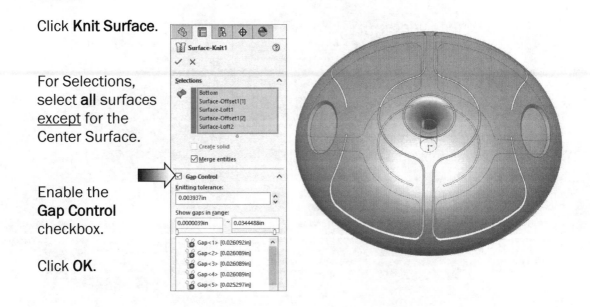

9. Adding fillets:

Click **Fillet**.

For Fillet Type, use the default **Constant Size**.

For Items to Fillet select **4 edges** of the two lofted features.

For Radius, enter **.030in**.

Click **OK**.

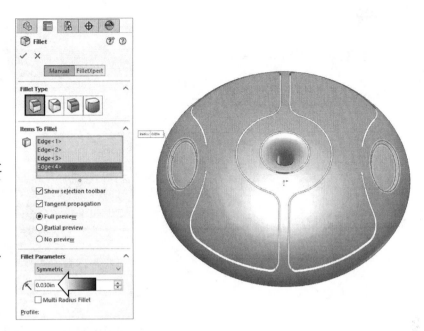

10. Mirroring a surface body:

Switch to the **Features** tab and click **Mirror**.

For Mirror Face/Plane select **Top** plane.

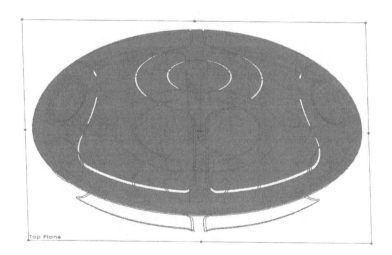

For Bodies to Mirror, select the **surface model** from the graphics area.

Click **OK**.

11. Separating the surface bodies:

Show the **Center Body**.

Select **Insert, Features, Move/Copy Bodies**.

For Bodies to Move, select the **Upper Body** and the **Center Body**.

For Distance, enter **3.00in** in the **Delta Y** direction.

Click **OK**.

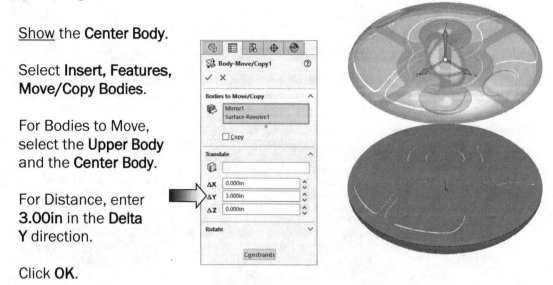

Repeat the step and move the Bottom Body, also **3.00in** along the -Y direction.

12. Calculating the surface Area:

Switch to the **Evaluate** tab.

Click **Measure**.

Select the **Upper Body** and enter the total surface <u>Area</u> here: _____ in ^2 (1 surface).

Enter the total surface <u>Area</u> of the **Lower-Body** here (13 surfaces): _____ in ^2

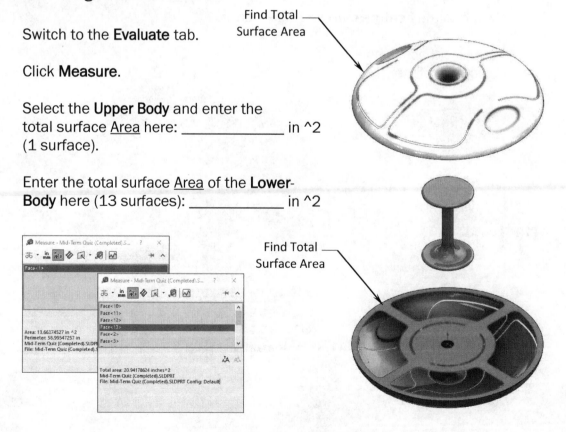

Chapter 7: Using Trim, Thicken & Configurations

Modem Housing

1. Opening a part document:

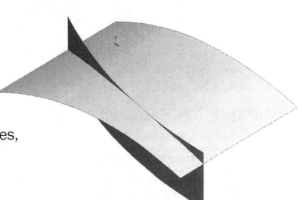

Open a part document named:
Modem Housing.sldprt

This document contains two extruded surfaces, two 2D-Sketches, and three 3D-Sketches.

2. Trimming the surfaces:

Change to the **Surfaces** tab.

Click **Trim Surface**.

For Trim Type, select **Mutual**.

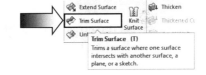

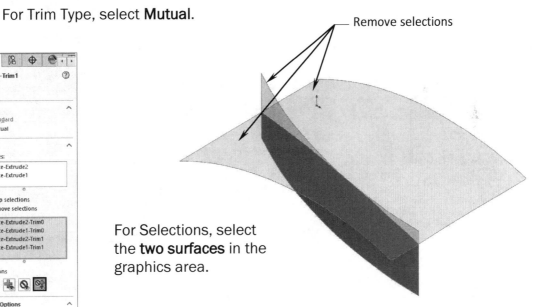

Remove selections

For Selections, select the **two surfaces** in the graphics area.

For Remove Selections, select the **three portions** of the surfaces as noted above.

Click **OK**.

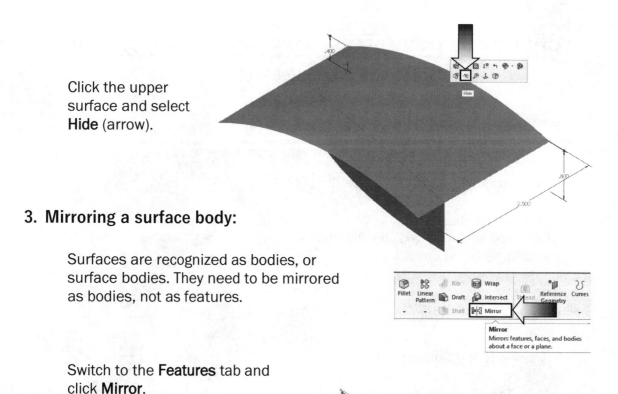

Click the upper surface and select **Hide** (arrow).

3. Mirroring a surface body:

Surfaces are recognized as bodies, or surface bodies. They need to be mirrored as bodies, not as features.

Switch to the **Features** tab and click **Mirror**.

For Mirror Face/ Plane, select the **Front** plane from the FeatureManager tree.

For Bodies to Mirror, select the **Trim Surface** from the graphics area.

Keep all other parameters at their default values.

Click **OK**.

<u>Show</u> the **five sketches** on the FeatureManager tree as noted below.

(Optional: Edit the sketches and add relations if needed.)

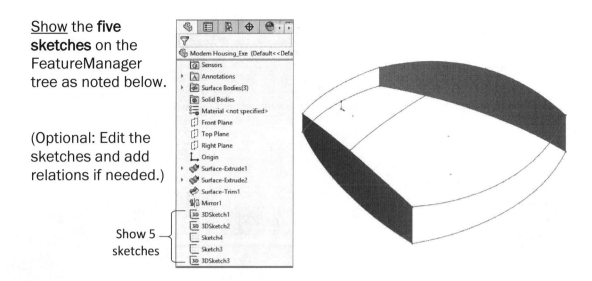

Show 5 sketches

4. Creating a lofted surface:

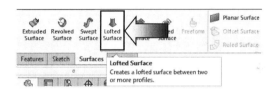

Switch to the **Surfaces** tab.

Click **Lofted Surface**.

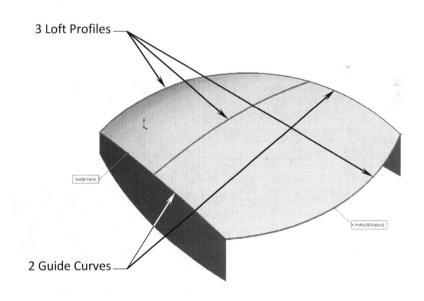

3 Loft Profiles

2 Guide Curves

Select the **3 Loft Profiles** and the **2 Guide Curves** as indicated.

Click **OK**.

5. Patching up the two ends:

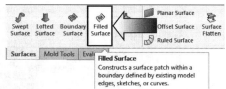

Click **Filled Surface**.

For Patch Boundary, select the **4 edges** on the right end.

For Curvature Control, select **Contact**.

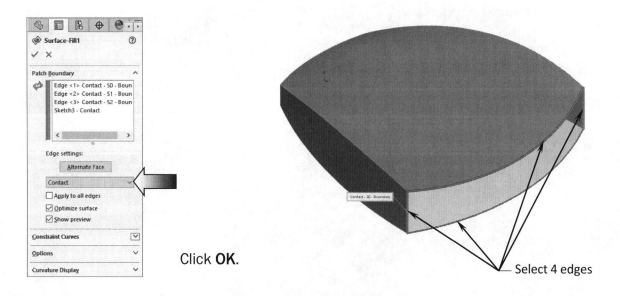

Click **OK.**

Select 4 edges

Click **Filled Surface** again.

Rotate the model to the opposite side and select the **4 edges** on the left end as indicated below.

Keep the Curvature Control at **Contact** to match the last one.

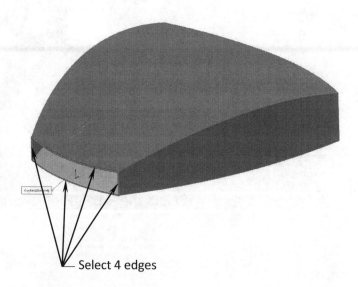

Click **OK.**

Select 4 edges

6. Creating the LED holes:

Open a **new sketch** on the <u>Right</u> plane.

Sketch **3 Circles** and a **Centerline.** Mirror 2 circles. There should be a total of 5 circles. Add a Horizontal relation between the centers of the circles.

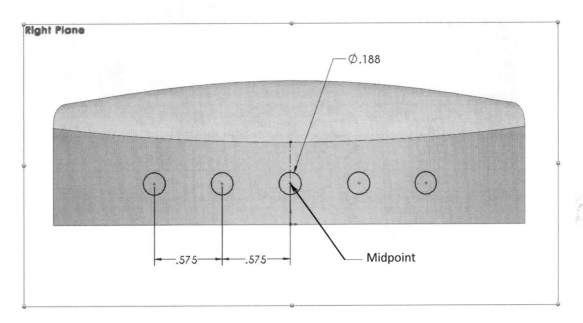

Add dimensions and the relation to fully define the sketch.

Switch to the **Surfaces** tab and click **Trim Surface**.

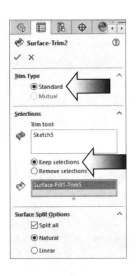

For Trim Type, click **Standard**.

For Trim Tool, select **Sketch5** (the 5 circles).

For Keep Selections, select the **outer portion** of the surface on the right end.

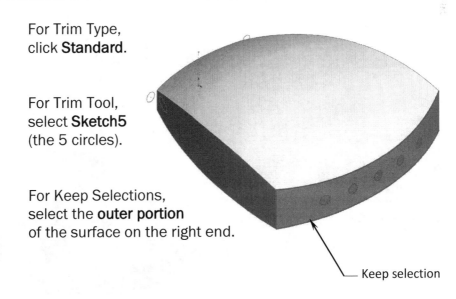

Keep selection

Click **OK**.

7. Making the power cord opening:

Select the <u>Right</u> plane and open a **new sketch**.

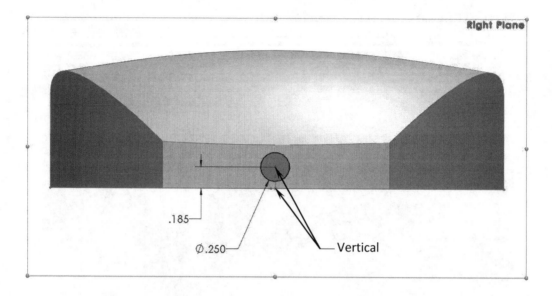

Sketch a **Circle** and add the dimensions and the relation shown above.

Switch to the **Surfaces** tab and click **Trim Surface**.

For Trim Type, click **Standard**.

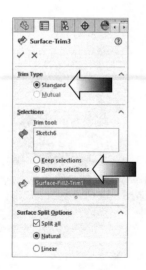

For Trim Tool, select **Sketch6** (the circle).

For Remove Selections, select the **inner portion** of the surface on left end, as noted.

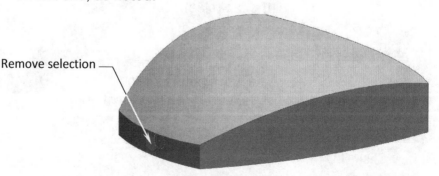

Remove selection

Click **OK**.

8. Knitting the surfaces:

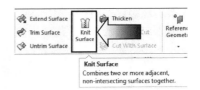

Click **Knit Surface**.

Knit Surface
Combines two or more adjacent,
non-intersecting surfaces together.

For Knit Selections, select all **5 surfaces** in the graphics area.

Enable the **Merge Entities** checkbox.

Click the **Gap Control** check-box and <u>enable</u> the checkboxes to allow the gaps to be healed.

Click **OK**.

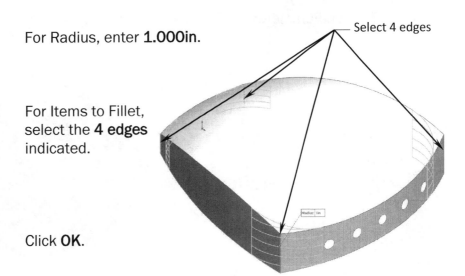

9. Adding the 1.00" fillets:

Click **Fillet**.

For Fillet Type, use the default **Constant Size** radius option.

For Radius, enter **1.000in**.

For Items to Fillet, select the **4 edges** indicated.

Click **OK**.

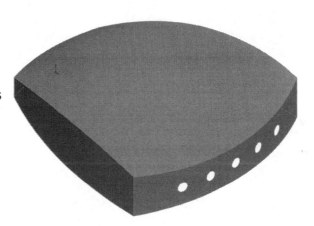

Select 4 edges

10. Adding the .125" fillets:

Click **Fillet** once again.

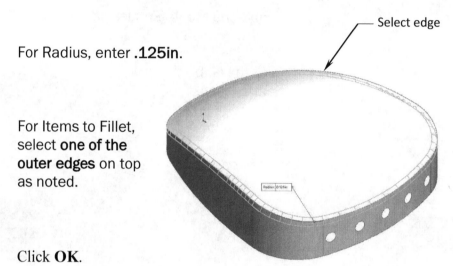

Use the default **Constant Size** radius option.

Select edge

For Radius, enter **.125in**.

For Items to Fillet, select **one of the outer edges** on top as noted.

Radius: 0.125in

Click **OK**.

11. Adding thickness:

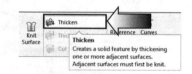

Knit Surface   Thicken   Reference Curves

Thicken
Creates a solid feature by thickening one or more adjacent surfaces.
Adjacent surfaces must first be knit.

Click **Thicken**.

For Thicken Parameters, select the **surface model** from the graphics area.

For Thickness location, select **Thicken Side 1** (Inside).

For Thickness enter: **.075in**.

Click **OK**.

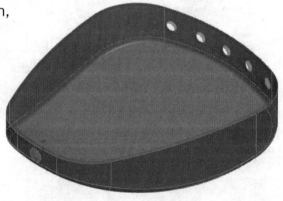

12. Creating the 1ˢᵗ recess cut:

Rotate the model to its bottom side.

Select the <u>bottom face</u> and open a **new sketch**.

Right click <u>one of the inner edges</u> and pick: **Select Tangency**.

Click **Convert Entities**.

All inner edges are converted into new sketch entities.

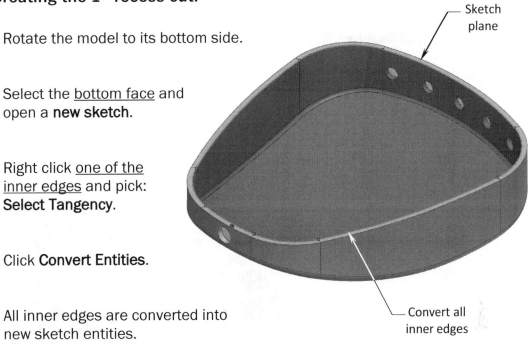

Sketch plane

Convert all inner edges

Right click <u>one of the converted entities</u> and pick: **Select Chain**.

Click **Offset Entities**.

For Offset Distance, enter **.035in**.

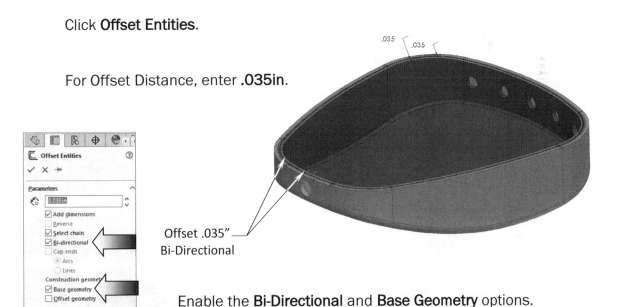

.035   .035

Offset .035"
Bi-Directional

Enable the **Bi-Directional** and **Base Geometry** options.

Click **OK**.

13. Making a cut:

Switch to the **Features** tab and click **Extruded Cut**.

For Direction 1, select the **Mid Plane** type.
(When cutting along curves, SOLIDWORKS sometimes displays the message "Zero Thickness Geometry". This was the reason we created a Bi-Directional offset in step number 12.)

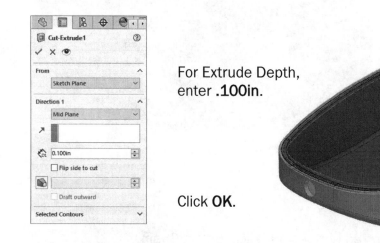

For Extrude Depth, enter **.100in**.

Click **OK**.

14. Creating the cover plate:

Open a **new sketch** on the bottom face.

Right-click one of the inner edges and pick: **Select Tangency**.

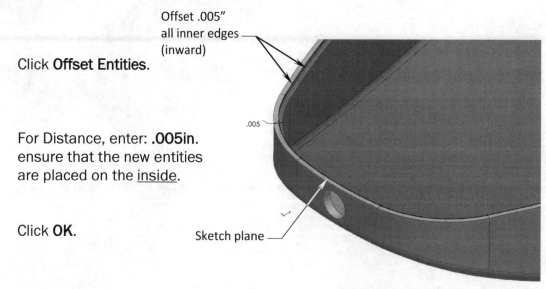

Offset .005"
all inner edges
(inward)

Click **Offset Entities**.

.005

For Distance, enter: **.005in**. ensure that the new entities are placed on the inside.

Click **OK**.

Sketch plane

Switch to the **Features** tab.

Click **Extruded Boss-Base**.

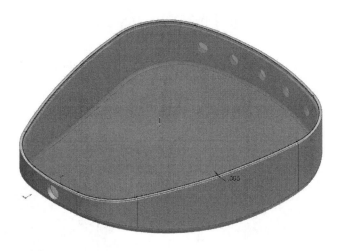

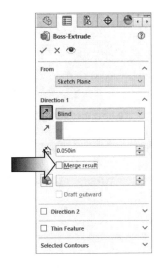

For Direction 1, use
the default **Blind** type.

For Extrude Depth, enter **.05in**.

Click **Reverse** and zoom closer to verify the extrude direction is downward, into
the part.

Clear the **Merge Result** checkbox to make
this extrude feature a new solid body.

Click **OK**.

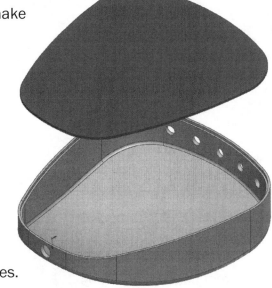

The exploded view shown on the
right is an example of the 2 solid bodies.
You do not need to create it just yet.

15. Creating the 2nd recess cut:

Open a **new sketch** on the <u>Top</u> plane.

Sketch the profile shown below.

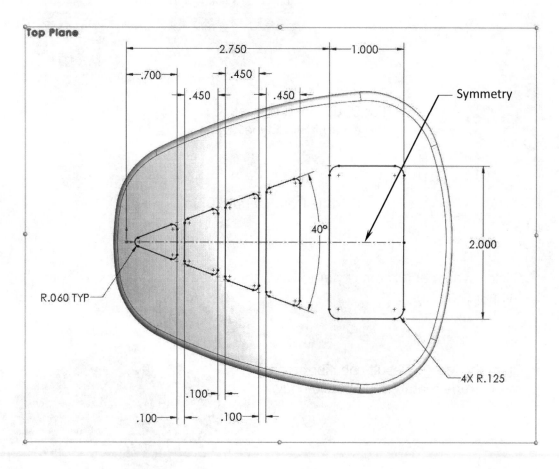

Add the dimensions / relations to fully define the sketch.

Add the Symmetry relations needed to eliminate the redundant dimensions.

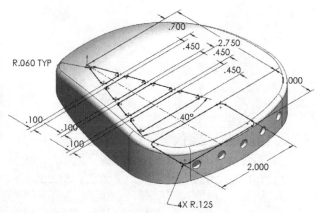

Switch to the **Features** tab.

Click **Extruded Cut**.

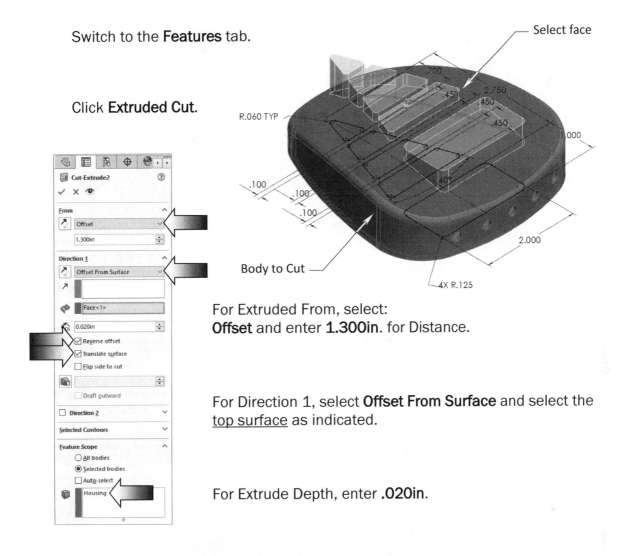

Select face

R.060 TYP

.100

.100

.100

Body to Cut

4X R.125

.700

.450

2.750

.450

.450

40°

1.000

2.000

For Extruded From, select:
Offset and enter **1.300in**. for Distance.

For Direction 1, select **Offset From Surface** and select the
<u>top surface</u> as indicated.

For Extrude Depth, enter **.020in**.

Enable the options: **Reverse Offset** and **Translate Surface**.

Expand the **Feature Scope**
section and select the
Housing Body to cut.

Click **OK**.

Inspect your model against the image
shown here.

16. Saving your work:

Select **File, Save As**.

Enter **Modem Housing_Completed.sldprt** for the file name.

Click **Save**.

Optionally, create an exploded view similar to the one shown above.
(Click: Insert, Exploded View. Select the upper Housing and drag the vertical arrow
to move along the Y direction.)

Using Configurations

Configurations creates multiple variations of a part or assembly model and saves them within a single document. Configurations are a convenient way to develop and manage families of models with different sizes, locations, components, materials, or visibility of components.

1. Opening a part document:

Open a part document named: **Configurations.sldprt**

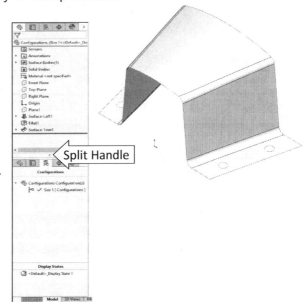

Drag the FeatureManager tree **Split-Handle** down about halfway.

The Default configuration has been renamed to **Size 1**.

2. Adding a new configuration:

Right-click inside the ConfigurationManager area and select: **Add Configuration** (arrow below).

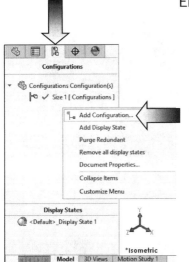

Enter: **Size 2** for Configuration Name (arrow).

Click **OK**.

A new configuration is created. Any changes done at this point will be captured and saved in the active configuration.

We will change the overall length and height dimensions of the part. The right end was constrained to update itself when the left side changes.

3. Modifying dimensions:

Double-click on **Plane1** from the FeatureManager tree.

Change 2.000 to 3.000

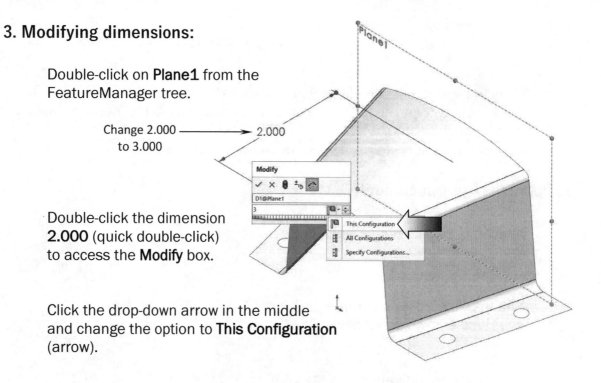

Double-click the dimension **2.000** (quick double-click) to access the **Modify** box.

Click the drop-down arrow in the middle and change the option to **This Configuration** (arrow).

Change the dimension **2.000** to **3.000** as indicated.

Expand the **Surface-Lofted1** and double click the **Sketch1** to display its dimensions.

Change 1.500 to 1.750

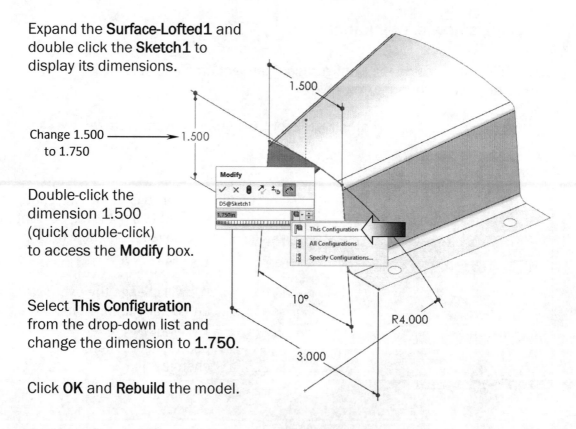

Double-click the dimension 1.500 (quick double-click) to access the **Modify** box.

Select **This Configuration** from the drop-down list and change the dimension to **1.750**.

Click **OK** and **Rebuild** the model.

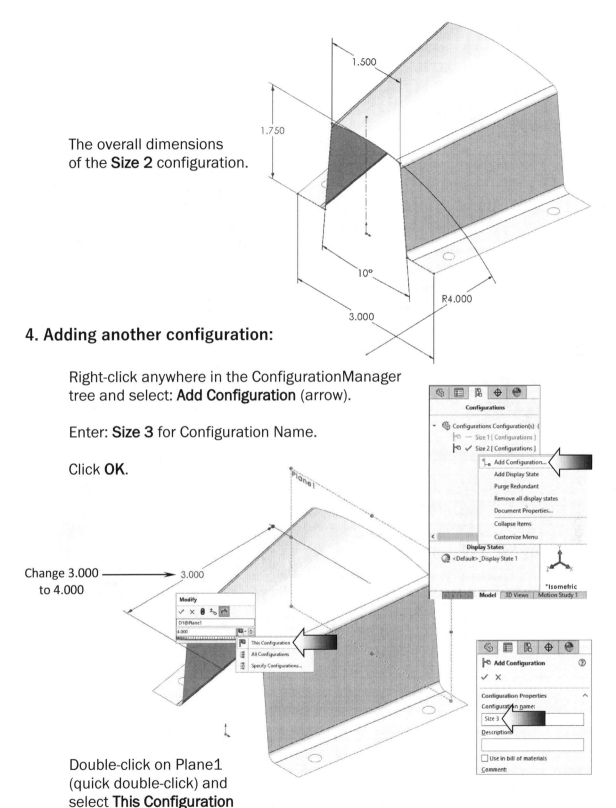

The overall dimensions of the **Size 2** configuration.

4. Adding another configuration:

Right-click anywhere in the ConfigurationManager tree and select: **Add Configuration** (arrow).

Enter: **Size 3** for Configuration Name.

Click **OK**.

Change 3.000 to 4.000

Double-click on Plane1 (quick double-click) and select **This Configuration** option from the drop-down list. Change the dimension **3.000** to **4.000**. Click **OK**.

Also change the height
dimension in the **Sketch1**
from **1.750** to **2.000**.

Change 1.750 ⟶ to 2.000

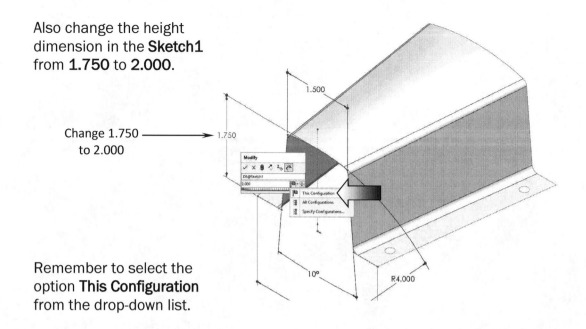

Remember to select the
option **This Configuration**
from the drop-down list.

Click **OK** and press the "stop light" to **Rebuild** the model.

There should be a total
of 3 configurations listed
on the Configuration-
Manager tree.

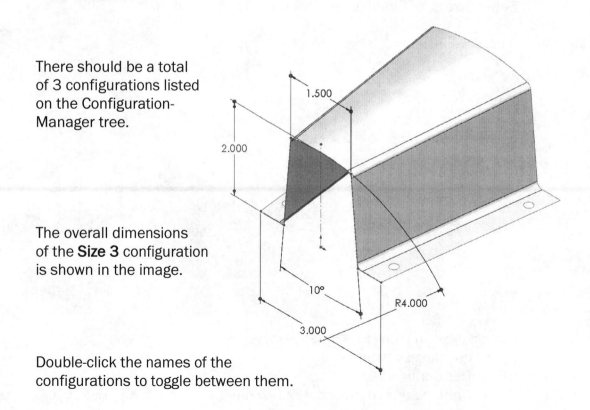

The overall dimensions
of the **Size 3** configuration
is shown in the image.

Double-click the names of the
configurations to toggle between them.

Optional:

*Add another configuration named **Size 4**.*

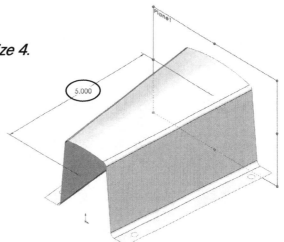

*Change the distance of **Plane1** to **5.000** (circled).*

*Change the overall height dimension of **Sketch1** from **2.000** to **2.250** (circled).*

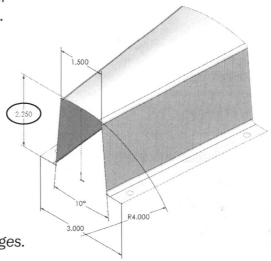

*Remember to set the option in the Modify box to **This Configuration**.*

***Rebuild** the model to update the changes.*

5. Saving your work:

Select **File, Save As**.

Enter: **Configurations_Completed.sldprt** for the file name.

Click **Save**.

Close all documents.

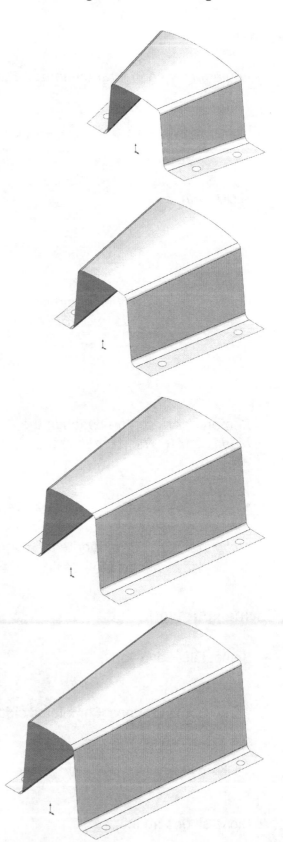

Configuration Size 1:

Height 1.500 X 2.000 Long

Configuration Size 2:

1.750 Height X 3.000 Long

Configuration Size 3:

2.000 Height X 4.000 Long

Configuration Size 4:

2.250 Height X 5.000 Long

Exercise: Trim, Knit & Patch Surfaces

The exercises give you an opportunity to apply what you have learned from the lessons. The instructions are limited to allow you to try things out using your own techniques, at your own pace.

1. Opening a part document:

Browse to the Training Folder and open a part document named: **Lesson 7_Exercise.sldprt**.

Click **No** to close the **Import Diagnostics** and also the **Feature Recognition** utilities.

There are 2 surfaces in this model.
They will be used to trim one another
and then knitted into a single surface body.

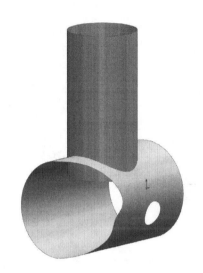

2. Trimming the surfaces:

The 2 surfaces are intersecting each other at this point.
The intersection portions can be removed with the trim tool.

Click **Trim Surface**.

For Trim Type, select
Mutual.

For Trimming Surfaces,
select the **2 surfaces**
in the graphics area.

For Remove Selections,
select the **2 inner portions**
of the surfaces as noted.

Click **OK**.

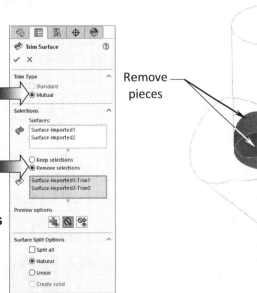

Remove pieces

3. Deleting the hole:

Instead of patching up the hole, the Delete Hole option can delete and fill the opening a little quicker.

Select the <u>edge</u> of the hole and press the **Delete** key on the keyboard.

The Choose Option dialog box pops up; click **Delete Hole(s)**.

Click **OK**.

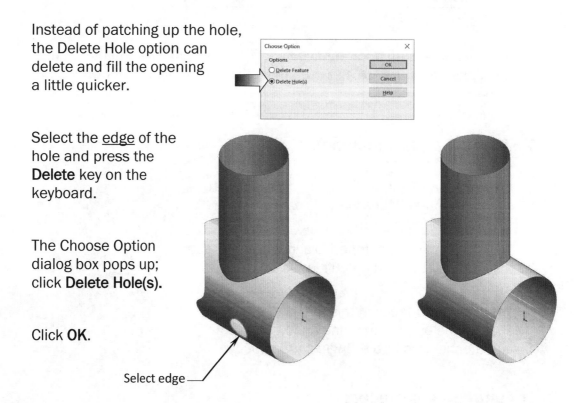

Select edge ——

4. Patching up the broken surface:

There are several ways to create the patch, but for this exercise we will take a look at the Filled Surface and the Extruded Surfaces options.

Open a new **3D Sketch**, below the Sketch button.

Select the <u>circular edge</u> on the right and click **Convert Entities**.

Select the Converted Circle and <u>delete</u> the **On Edge** relation. This will allow the circle to move to a different location.

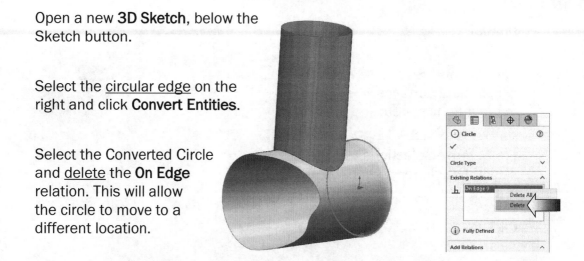

Hold the <u>Control</u> key and select the converted **Circle** and the circular **Edge** as indicated.

Click **Coradial**. This relation causes the converted circle to snap to the circular edge and share the same diameter.

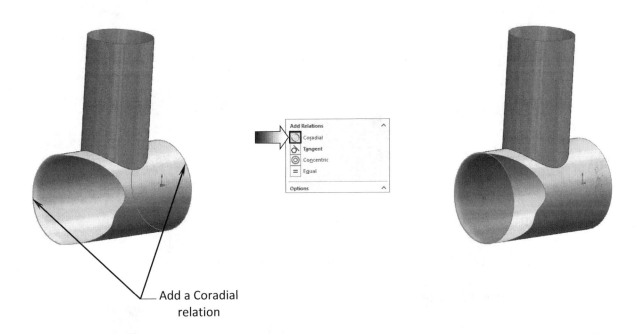

Add a Coradial
relation

Add a vertical **centerline** and **trim** away the left portion of the circle.

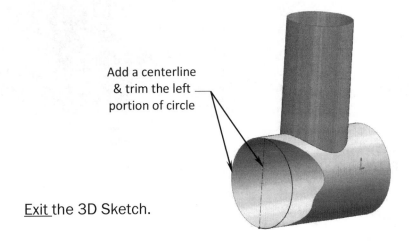

Add a centerline
& trim the left
portion of circle

<u>Exit</u> the 3D Sketch.

5. Trying a Filled Surface:

Click **Filled Surface**.

For Patch Boundary, select the **3 edges** of the broken area and the **3D Sketch**.

Select the **3 edges** from the dialog box and change the Curvature Control to **Tangent**.

Click **OK**.

The Filled Surface tool does not produce a high quality surface that we are looking for. Undo the Filled Surface feature.

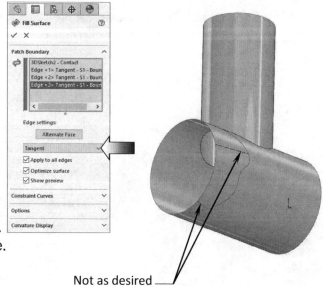

Not as desired

6. Creating an Extruded Surface:

Select the **3D Sketch** from the FeatureManager tree.

Click **Extruded Surface**.

Extruding a 3D Sketch will require a direction.

Click **View, Hide/Show, Temporary Axis.**

For Direction 1, select **Up-to-Surface** and click **Reverse**.

For Direction of Extrusion, select the **Vertical Temporary Axis.**

For Face/Plane, select the **lower-horizontal surface body.**

Click **OK.**

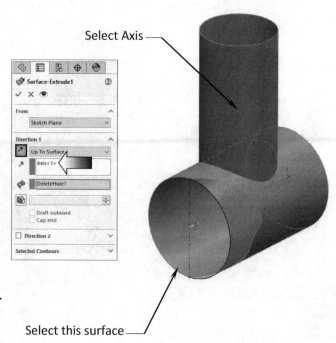

Select Axis

Select this surface

7. Knitting the surfaces:

The extruded surface looks much better than the filled surface.
The edges between the patch and the horizontal surface body will be removed after the knit operation.

Click **Knit Surface**.

For Surfaces to Knit, select the **2 surface bodies** in the graphics area.

Enable the **Merge Entities** check box.

Click **OK**.

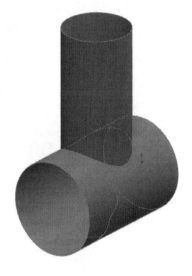

8. Saving your work:

Select **File, Save As**.

Enter **Lesson 7_Exercise_Completed** for the file name.

Click **Save**.

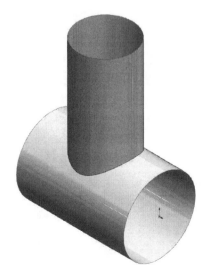

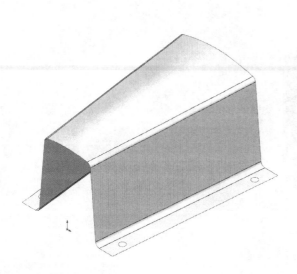

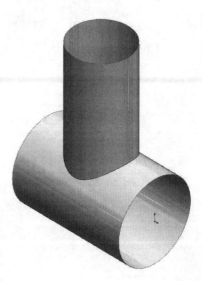

Chapter 8: Curved Driven Pattern & Flex Bending

Stent Design Examples

1. Opening a part document:

Open a part document named:
Curve Driven Pattern.sldprt

There are 3 surfaces in this document: two swept surfaces and an extruded surface. The extruded surface needs to be replicated 12 times along the spiral curve. The Curve Driven Pattern option will be used to copy the extruded surface.

2. Creating a 3D sweep path:

Select **3D-Sketch** under the Sketch drop-down menu (arrow).

Select the **curve** that connects to the line and click: **Convert Entities**.

<u>Drag</u> the upper end point of the curve downward to shorten it.

Add the spacing dimension .200in from the <u>end point</u> of the curve to the <u>circular edge</u> of the spiral surface.

<u>Exit</u> the 3D-Sketch.

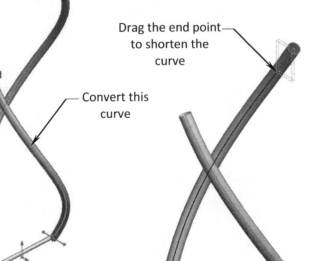

Drag the end point to shorten the curve

Convert this curve

.200

3. Creating a swept surface:

One of the requirements when using the Curve Driven Pattern option is to have a reference surface for the pattern instances to be aligned normal to. The next couple of steps will walk us through the creation of this reference surface.

Click **Swept Surface** on the **Surfaces** tab.

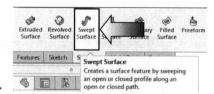

For Sketch Profile, select the **Line** at the bottom.

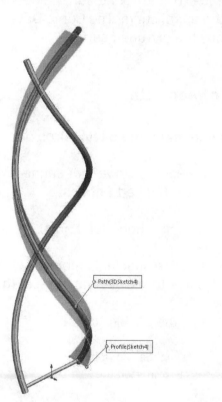

For Sweep Path, select the **3-D Sketch Curve**.

The preview graphics show the line is being swept along the path.

Click **OK**.

4. Creating a Curve Driven Pattern:

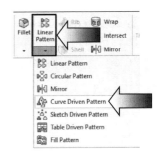

Switch to the **Features** tab.

Select the **Curve Driven Pattern** command under the Linear Pattern drop-down menu.

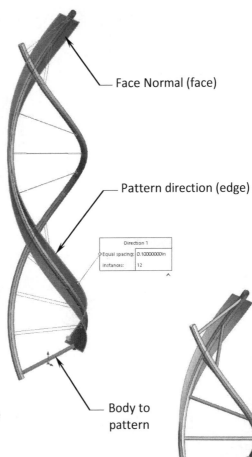

For Pattern Direction select the **Edge** of the swept surface as noted.

Face Normal (face)

Pattern direction (edge)

For Number of Instances, enter **12**.

Under Curve Method, select the following:

       * **Offset Curve**
       * **Tangent To Curve**

Body to pattern

Under Face Normal, select the **Swept Surface** as indicated.

Enable the <u>Bodies</u> checkbox and select the **Extruded Surface** as Body to Pattern.

Click **OK**.

5. Saving your work:

Hide the Surface Swept1.

Select **File, Save As**.

Enter: **Curve Driven Pattern_Completed.sldprt** for the file name.

Click **Save**.

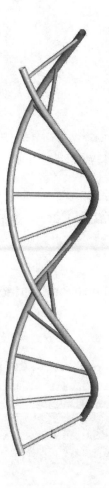

Stent Designs – Exercise 1

1. Opening a part document:

Open a part document named:
Stent 1.sldprt

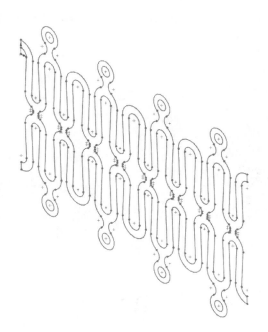

A 2D sketch pattern has already been created to help us focus on making a 3D rolled pattern using the Flex-Bending option.

2. Creating a Planar Surface:

Switch to the **Surfaces** tool tab.

Select the **Planar Surface** command (arrow).

For Bounding-Entities, select **Sketch1** either from the FeatureManager tree or select it directly in the graphics area.

Click **OK**. A planar surface is created from a 2D sketch.

3. Creating a Rolled Pattern:

Select **Insert, Features, Flex** (arrow).

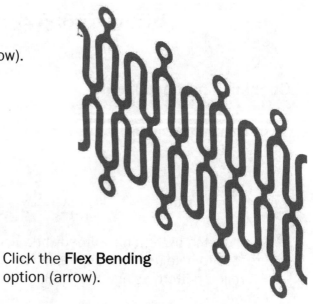

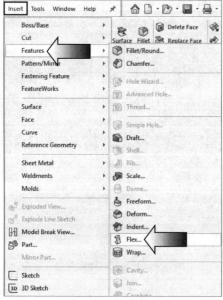

Click the **Flex Bending** option (arrow).

For Bend Angle, enter **360°** (arrow).

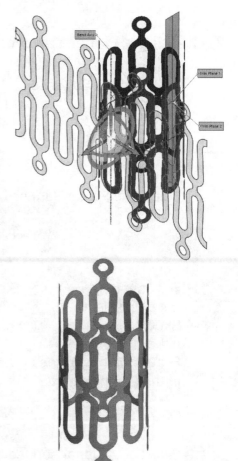

Set the Rotation-Angle as follows:

X Rotation **270deg**

Y Rotation **90deg**

Z Rotation **0deg**

Click **OK**.

The rolled pattern is created from a flat-pattern.

4. Thickening the surface model:

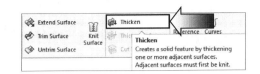

Select the **Thicken** command (arrow) from the **Surfaces** tab.

For Thicken Parameters, select the **surface model** (Flex1) directly from the graphics area.

For Thickness location, select the **Thicken Both Sides** option (arrow).

For Thickness, enter **.006in** (or .012in total).

Zoom in closer to inspect the thickness. The preview graphics should show the thickness is being added to <u>both sides</u> of the surface model.

Click **OK**.

5. Saving your work:

Save your work as:

Stent 1_Completed.sldprt

Stent Designs – Exercise 2

1. Starting a part document:

Select **File, New, Part**.

Change the Units to **IPS** and the number of decimals to **3 places**.

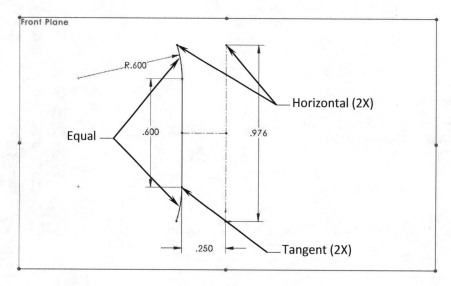

Open a **new sketch** on the <u>Front</u> plane.

Sketch the profile and add the dimensions / relations as shown in the image.

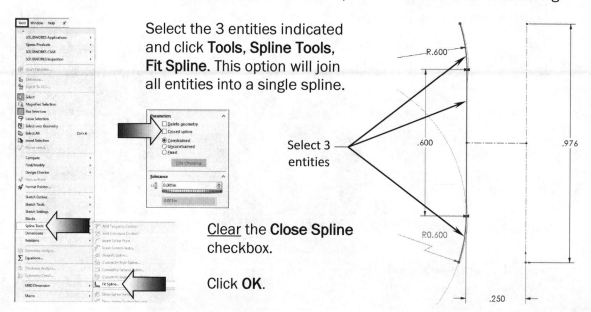

Select the 3 entities indicated and click **Tools, Spline Tools, Fit Spline**. This option will join all entities into a single spline.

<u>Clear</u> the **Close Spline** checkbox.

Click **OK**.

Switch to the **Surfaces** tab and click **Revolved-Surface**.

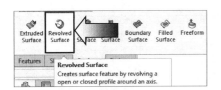

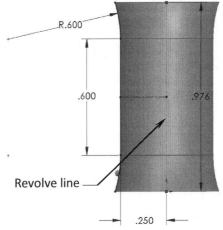

Select the **revolve centerline** and click **OK**.

2. Creating a Split Sketch:

Open another **sketch** on the Front plane.

Sketch the profile shown below. Use the mirror option to keep the sketch entities symmetric with each other about the vertical centerline.

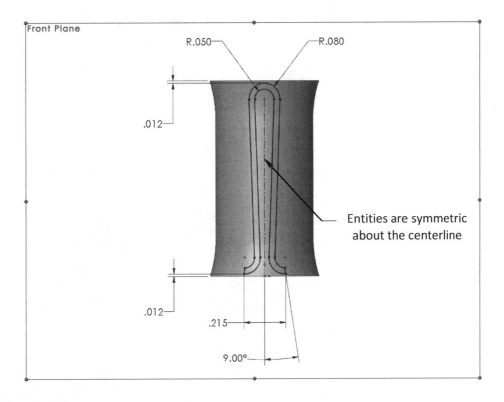

Add the dimensions / relations shown above to fully define the sketch.

3. Creating a Split Line feature:

Select **Curves, Split Line** on the **Surfaces** tab.

Use the default **Projection** type.

For Selections, select **Sketch1** and then click the **surface model**.

Enable the 2 checkboxes:
    Single Direction and
    Reverse Direction

Click **OK**.

4. Deleting faces:

The Delete Face command deletes one or more faces from a solid or surface body.

Select the **Delete Face** command (arrow).

For Selections, select the **Revolved Surface**.

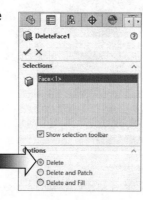

Under Options, select **Delete** (arrow).

Click **OK**.

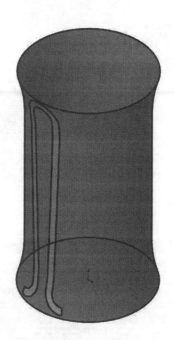

The resulting surface delete is shown here.

The selected portion of the surface body is deleted; only the inner portion is kept.

5. Adding thickness:

The Thicken command creates a solid feature by thickening one or more adjacent surfaces.

Click **Thicken** (arrow).

For Thicken Parameters, select the **surface model**.

For Thickness location, select: **Both Sides**.

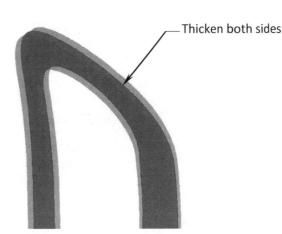

— Thicken both sides

For Thickness, enter **.005in** (.010in. total). Zoom closer to verify the thickness is being added to both sides of the surface model.

Click **OK**.

6. Creating an axis:

Either a circular edge or an axis is required as the center of the circular pattern. We will create an axis for this next step.

Select **Axis** from the **Reference Geometry** pull-down menu (arrow).

For Selections, select the **Two Planes** option (arrow).

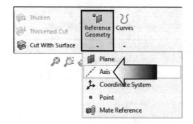

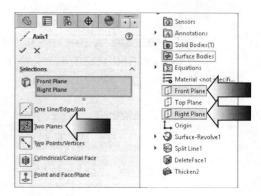

Expand the FeatureManager tree and select the **Front** and the **Right** planes.

Click **OK**.

The Axis may not be visible in the graphics area by default.

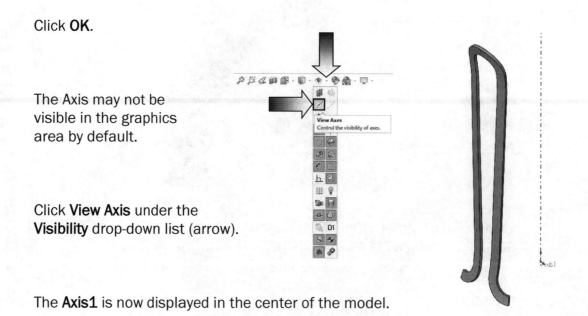

Click **View Axis** under the **Visibility** drop-down list (arrow).

The **Axis1** is now displayed in the center of the model.

7. Creating a circular pattern:

There will be a slight overlap between the instances. However, when combining the solid bodies, the overlaps will be removed automatically.

Switch to the **Features** tab.

Select the **Circular Pattern** command from the Linear Pattern drop-down list.

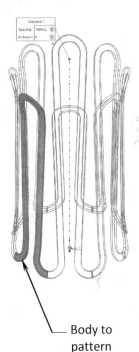

For Direction 1, select **Axis1**.

Click the **Equal Spacing** option.

Use the default **360°** and enter: **8** for total number of instances.

Expand the Bodies section and select the **solid body** as noted.

Body to pattern

The preview graphics show a pattern of 8 instances is being created.

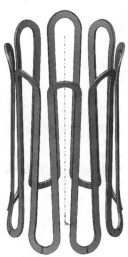

Click **OK**.

Notice a slight overlap in between each instance? Again, they will be removed when all instances are combined as one.

8. Combining the solid bodies:

Select **Insert, Features, Combine (arrow)**.

For Operation Type, select **Add**.

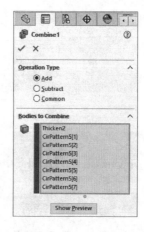

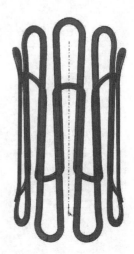

For Bodies to Combine, select all **8 solid bodies** either from the graphics area or in the Solid Bodies folder, on the FeatureManager tree.

Click **OK**.

9. Saving your work:

Select **File, Save As**.

Enter: **Stent 2_Completed.sldprt** for the file name.

Click **Save**.

Stent Designs – Exercise 3

1. Opening a part document:

Open a part document named: **Stent 3.sldprt.**

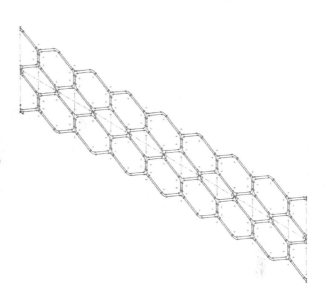

A 2D sketch has been created to help focus on the technique of making this unique stent.

2. Creating the base sketch:

Open a **new sketch** on the <u>Front</u> plane.

Sketch a **Corner Rectangle** starting from the Origin.

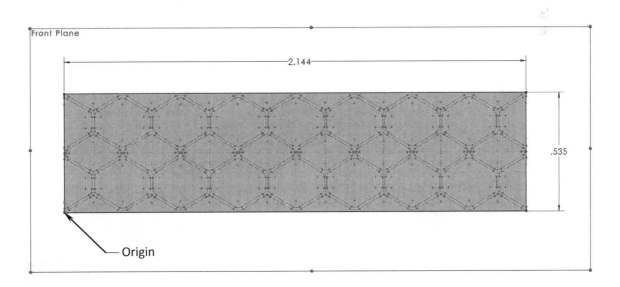

Add the width and height dimensions to fully define the sketch.

3. Creating a planar surface:

Switch to the **Surfaces** tab and click: **Planar Surface.**

For Bounding Entities, select the **rectangle.**

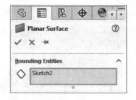

Click **OK.**

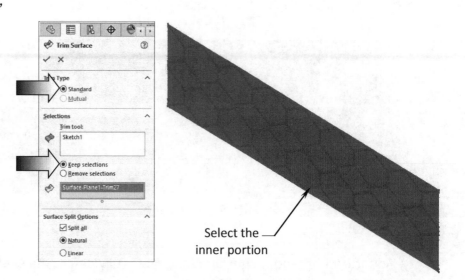

4. Trimming with a sketch:

Click **Trim Surface** (arrow).

The **Standard Trim** option is used when trimming with a sketch.

For Trim Tool, select the **Sketch1.**

For Keep Selections, select the **inner-portion** of the surface body.

Click **OK.**

Select the inner portion

5. Thickening the surface:

Click **Thicken** on the **Surfaces** tab.

For Thicken Parameters, select the **Surface Body** directly from the graphics area.

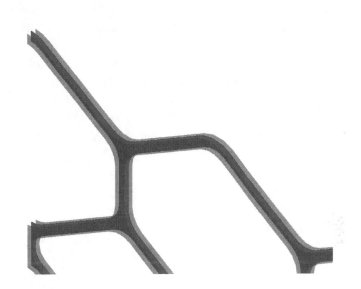

For Thickness, select **Both Sides**.

Zoom closer to verify the thickness is being added to both sides of the surface model.

For Thicken Thickness, enter: **.006in**. (or .012in total).

Click **OK**.

6. Creating a flex bending feature:

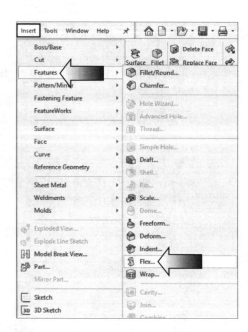

The last step is to bend the flat pattern into a round, cylindrical shape.

Select **Insert, Features, Flex** (arrow).

Click the **Bending** option (arrow).

For Bend Angle, enter **359.95°** (to be able to see where it joins).

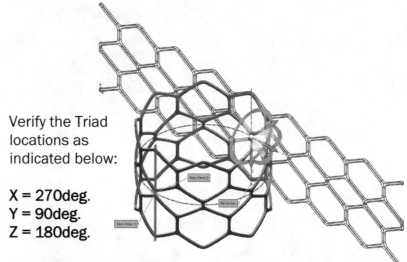

Verify the Triad locations as indicated below:

X = 270deg.
Y = 90deg.
Z = 180deg.

Click **OK**.

7. Saving your work:

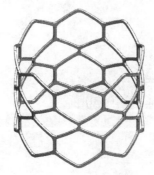

Select **File, Save As**.

Enter: **Stent 3_Completed** for the file name.

Click **Save**.

Exercise: Pattern & Flex Bending

The exercises give you an opportunity to apply what you have learned from the lessons. The instructions are limited to allow you to try things out using your own techniques, at your own pace.

1. Opening a part document:

Browse to the Training Folder and open a part document named: **Lesson 7_Exercise.sldprt**.

Click **No** to close the **Import Diagnostic** and also the **Feature Recognition** utilities.

There is only 1 surface in this model. It will get patterned 10 times and then twisted into its final shape.

2. Creating a Circular Pattern:

Switch to the **Features** tab.

Click **Linear Pattern**.

For Pattern Direction select the <u>vertical line</u> as noted.

For Spacing, enter **.062in**.

For Number of Instances, enter **10**.

For Bodies to Pattern select **Surface Trim1** in the graphics area.

Click **OK**.

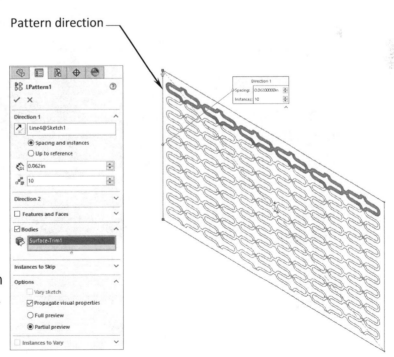

Pattern direction

3. Knitting the surfaces:

Switch back to the **Surfaces** tab.

Click **Knit Surface**.

For Knit Selections, select all <u>10 surfaces</u> from the graphics area.

Enable the **Merge Entities** checkbox.

Click **OK**.

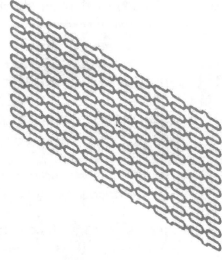

4. Creating a Flex Bending feature:

Select **Insert, Features, Flex**.

Use the default **Flex Bending** option.

For Flex Input select the **Knit Surface**.

For Bend Angle enter **180deg**.

Locate the Triad section; in the Y Rotation Angle, enter **30deg**.

Drag the Flex Accuracy to <u>max</u>.

Click **OK**.

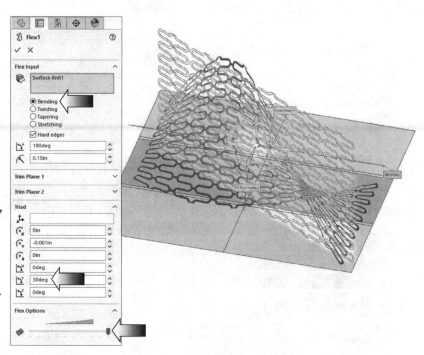

5. Saving your work:

Select **Files, Save As**.

Enter **Lesson 8_Exercise_Completed** for the file name.

Click **Save**.

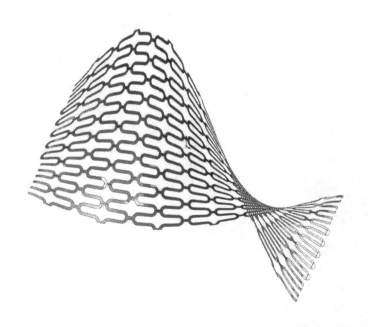

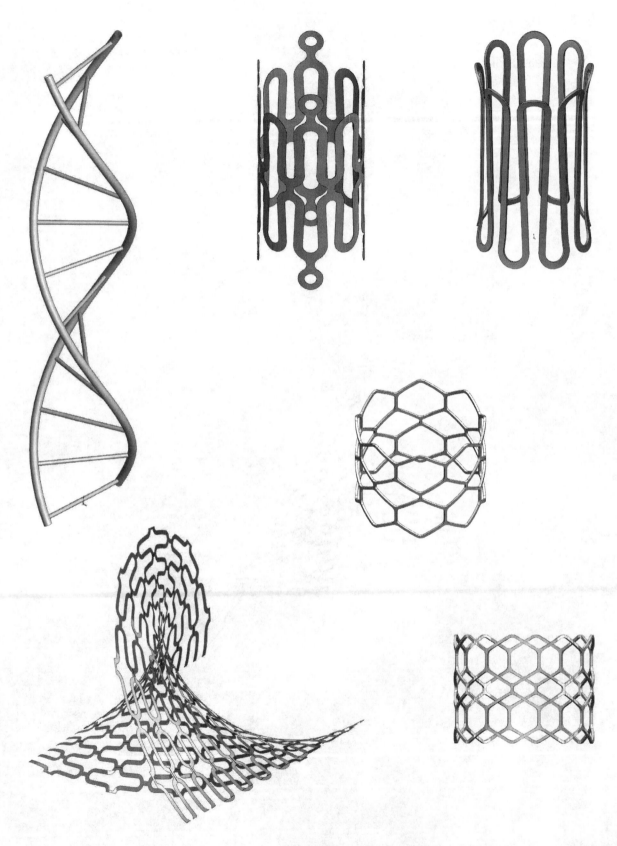

Exercise: Using Surfaces

The exercises give you an opportunity to apply what you have learned from the previous lessons. The instructions are limited to allow you to try things out using your own techniques, at your own pace.

1. Opening a part document:

Browse to the Training Folder and open a part document named: **Using Surfaces.sldprt**.

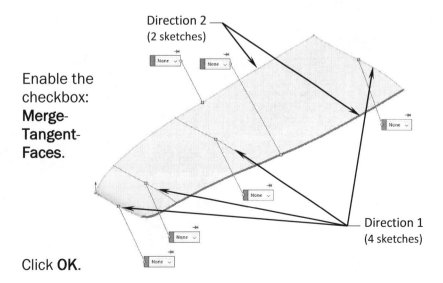

2. Creating a boundary surface:

Switch to the **Surfaces** tab and click **Boundary Surface**.

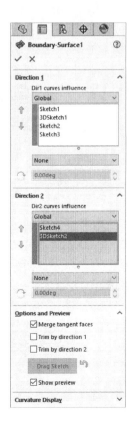

For Direction 1, select the **4 sketches** as indicated below (Sketch1, 3Dsketch1, Sketch2, and Sketch3).

For Direction 2, select the **2 sketches** from the Feature-Manager tree (Sketch4 and 3DSketch2).

Enable the checkbox: **Merge-Tangent-Faces**.

Click **OK**.

3. Mirroring a surface body:

Switch to the **Features** tab and click **Mirror**.

For Mirror Plane, select the **Right** plane.

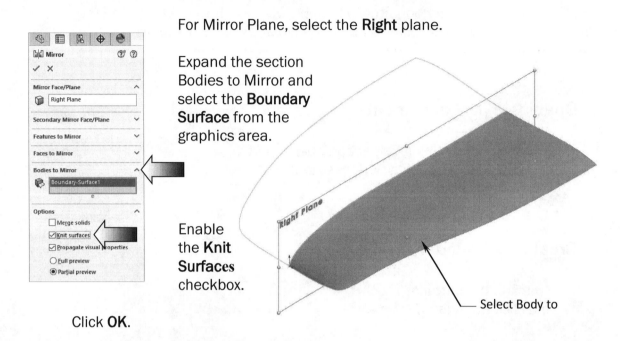

Expand the section Bodies to Mirror and select the **Boundary Surface** from the graphics area.

Enable the **Knit Surfaces** checkbox.

Select Body to

Click **OK**.

4. Adding an extruded surface:

Select **Sketch5** from the FeatureManager tree and click **Extruded Surface** on the **Surfaces** tab.

For Direction 1, select **Mid Plane**.

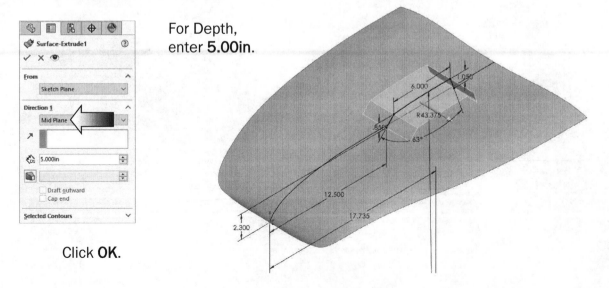

For Depth, enter **5.00in**.

Click **OK**.

5. Making a trim-sketch:

Expand the Surface Bodies folder, right-click the Mirror1 and select: **Change Transparency**.

Open a **new sketch** on the Top plane.

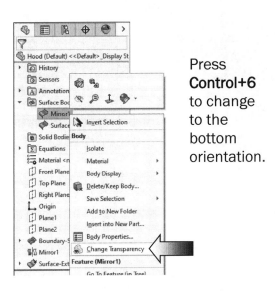

Press **Control+6** to change to the bottom orientation.

Sketch **2 lines** and add the dimensions shown in the image.

6. Extruding a surface:

Switch to the **Surfaces** tab and click **Extruded Surface**.

For Direction 1, select **Mid Plane** from the list.

For Depth, enter **5.000in**.

Click **OK**.

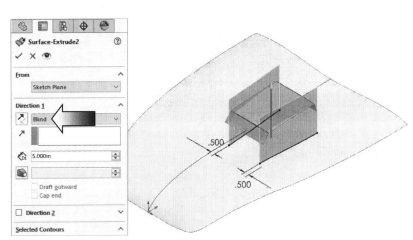

7. Trimming the surfaces:

Click **Trim Surface**.

For Trim Type, select **Mutual**.

For Selections, select all **4 surfaces**.

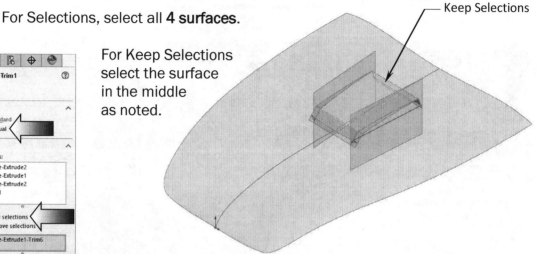

Keep Selections

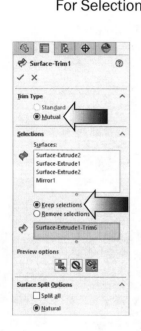

For Keep Selections select the surface in the middle as noted.

Toggle between the 3 preview options to see the include or exclude regions more clearly.

Click **OK**.

8. Hiding the surface bodies:

Using the Surface Bodies folder, Hide the 2 extruded surfaces.

We will try a different approach to patch up the 2 sides of the raised surface.

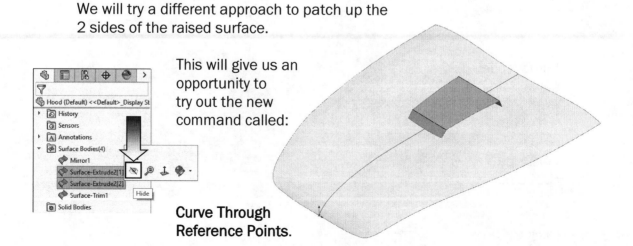

This will give us an opportunity to try out the new command called:

Curve Through Reference Points.

9. Creating the 1st reference curve:

Expand the Curves drop-down list and select:
Curve Through Reference Points.

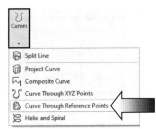

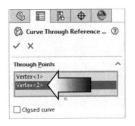

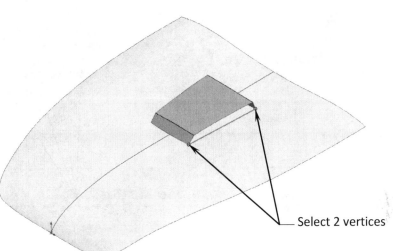

Select the **2 vertices** on the side of the raised surface.

The preview graphics show a new curve is being created.

Click **OK**.

Select 2 vertices

10. Creating the 2nd reference curve:

Click **Curve Through Reference Points** again.

Rotate the model and select the **2 vertices** on the other side of the raised surface.

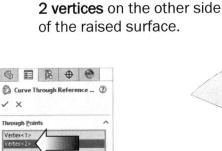

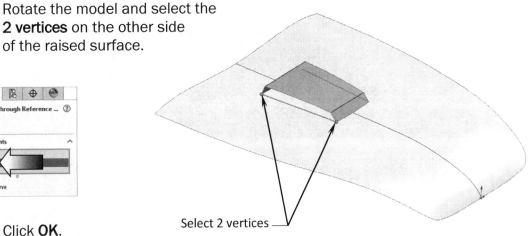

Click **OK**.

Select 2 vertices

11. Adding the 1st planar surface:

For planar openings, the Planar Surface command is used to create the patches.

Click **Planar Surface**.

For Bounding Entities, select the **3 edges** and **1 reference curve** as indicated.

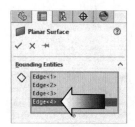

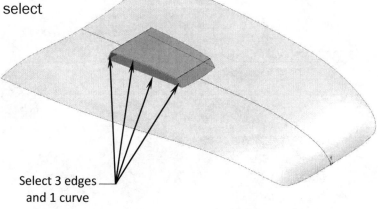

Select 3 edges and 1 curve

Click **OK**.

12. Adding the 2nd planar surface:

Click **Planar Surface** once again.

Rotate the model and select the **3 edges** and **1 reference curve** on the opposite side of the raised surface.

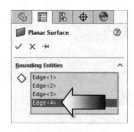

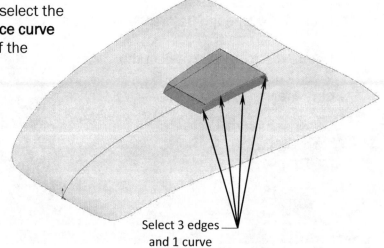

Select 3 edges and 1 curve

Click **OK**.

The 2 sides of the raised surface are patched up with 2 planar surfaces.

13. Knitting the surface bodies:

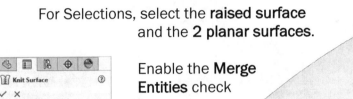

For clarity, <u>hide</u> the 2 reference curves (and optionally deselect the Transparency option of the main surface).

Click **Knit Surface**.

For Selections, select the **raised surface** and the **2 planar surfaces**.

Enable the **Merge Entities** check box.

Click **OK**.

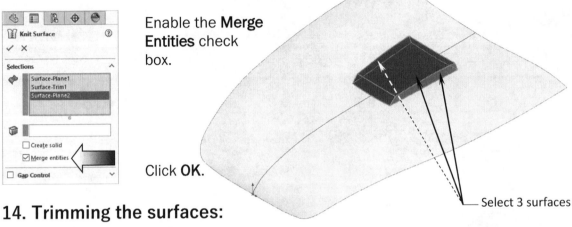

Select 3 surfaces

14. Trimming the surfaces:

Rotate the model so that the bottom of the raised surface is visible for the next trim operation.

Click **Trim Surface** once again.

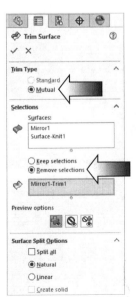

For Trim Type, select **Mutual**.

For Selections, select the **raised surface** and the **Knit surface**.

For Remove Selections, select the **middle portion** of the raised surface from the bottom, as indicated.

Click **OK**.

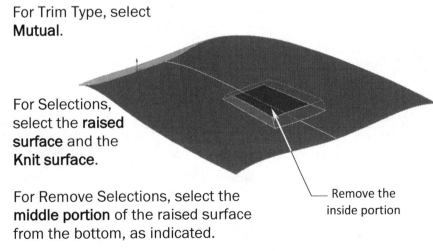

Remove the inside portion

15. Knitting the surfaces:

We need to knit the surfaces into a single surface so that fillets can be added to the surface model.

Click **Knit Surface** and select **both surfaces** in the graphics area.

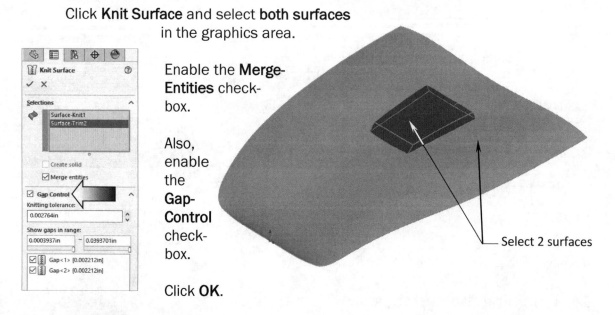

Enable the **Merge-Entities** check-box.

Also, enable the **Gap-Control** check-box.

Click **OK**.

Select 2 surfaces

16. Adding the .750" fillets:

Click **Fillet**. For Fillet Type, use the default **Constant Size** radius option.

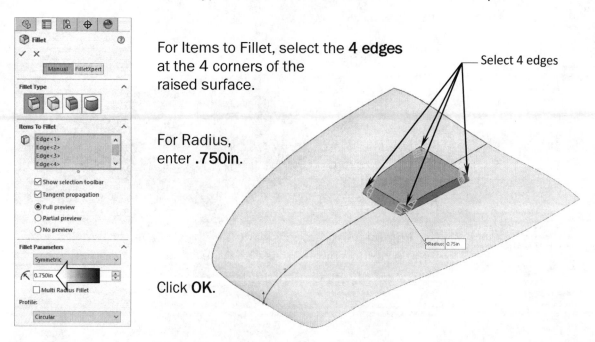

For Items to Fillet, select the **4 edges** at the 4 corners of the raised surface.

Select 4 edges

For Radius, enter **.750in**.

Click **OK**.

17. Adding the .500" fillets:

Click **Fillet** again and continue to use the default **Constant Size** radius option.

For Items to Fillet, right-click one of the edges at the bottom of the raised surface and pick: **Select Tangency**.

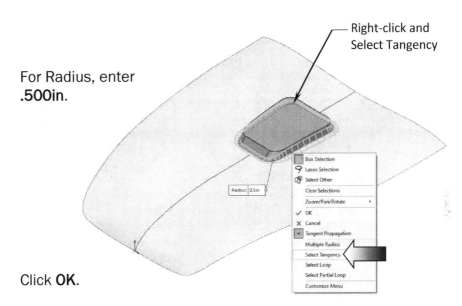

For Radius, enter **.500in**.

Right-click and Select Tangency

Click **OK**.

18. Adding the .250" fillets:

Click **Fillet** once again.

Using the same **Constant Size** radius option, select one of the **upper edges** of the raised surface as noted.

Enter **.250in** for radius.

Click **OK**.

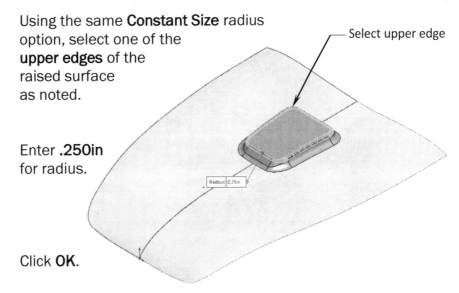

Select upper edge

19. Saving your work:

Select **File, Save As**.

Enter: **Using Surfaces_Completed.sldprt** for the file name.

Click **Save**.

Close all documents.

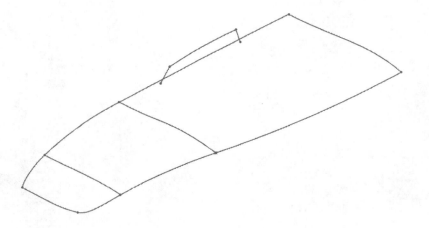

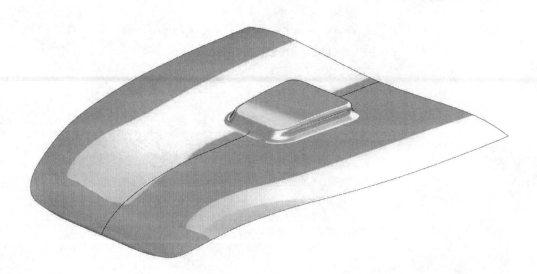

Chapter 9: Surfaces & Solids - Hybrid Modeling
Catheter Housing

This lesson explores the steps taken to accomplish hybrid modeling. Below, we create a Handle Housing to demonstrate the functionality of these tools.

1. Opening a part document:

Open a part document named: **Handle Housing.sldprt**

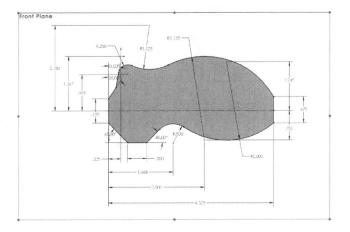

A 2D sketch has been created to use as the base sketch for this surface model.

2. Creating the 1st extruded surface:

Change to the **Surfaces** tab.

Click the **Extruded Surface** command (arrow).

Enter the following:

    * **Blind**

    * **Reverse**

    * **.625in**

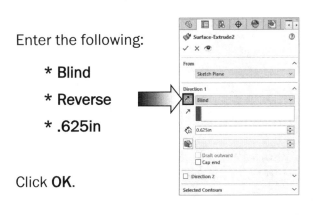

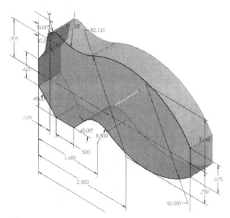

Click **OK**.

Zoom / rotate the surface model to view it from different angles.

Your extruded surface should look like the image shown here.

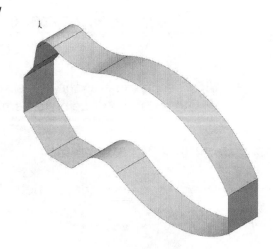

3. Creating the 2nd extruded surface:

Open a **new sketch** on the <u>Top</u> plane.

Sketch a **3-Point Arc** as indicated.

Add the dimensions shown in the image to fully define the sketch.

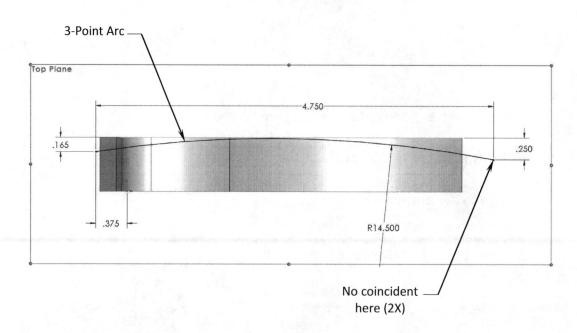

Note that the left and right ends of the arc are <u>not</u> coincident to the surface model.

The arc should clear the top edge of the model and there should be a small gap between the left end of the arc and the surface model.

Switch to the **Surfaces** tab.

Click **Extruded Surface**.

For Direction 1, use the default **Blind** type.

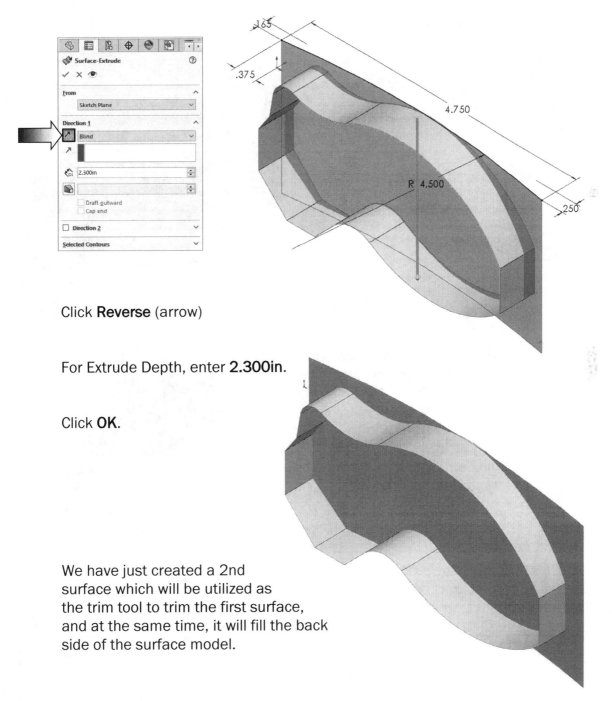

Click **Reverse** (arrow)

For Extrude Depth, enter **2.300in**.

Click **OK**.

We have just created a 2nd
surface which will be utilized as
the trim tool to trim the first surface,
and at the same time, it will fill the back
side of the surface model.

4. Trimming the surfaces:

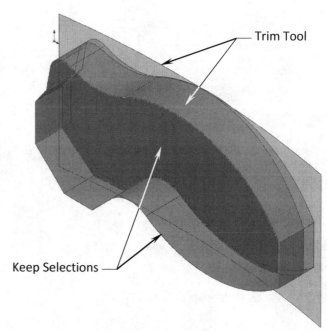

Trim Tool

Click **Trim Surface** on the **Surfaces** tab.

For Trim Type, select the **Mutual** trim option.

For Trim Tool, select both, the **1st** and **2nd Extruded Surfaces.**

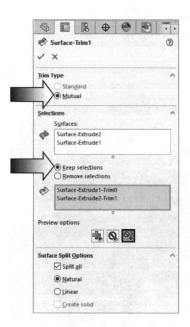

Keep Selections

Click the **Keep Selections** button and select the front portion of the **1st Extruded Surface** and the inner portion of the **2nd Extruded Surface** as indicated.

Under the Surface Split Options, enable the **Split All** and **Natural options.**

Click **OK.**

Helpful Tips

A surface, plane, or sketch can be used as a trim tool to trim intersecting surfaces. Additionally, a surface can be used in conjunction with additional surfaces, such as mutual trim.

* **Natural:** Boundaries extend tangent from the ends of the Trim tool.
* **Linear:** Boundaries extend from the Trim tool endpoints to the nearest edge.

5. Adding the .500" fillets:

Select the **Fillet** command from the **Surfaces** tab.

For Fillet Type, use the default **Constant Size** option.

For Fillet Size, enter **.500in**.

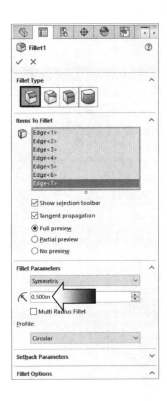

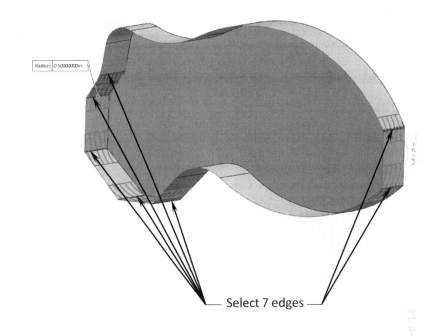

Select 7 edges

For Items to Fillet, select the **7 edges** as indicated.

Enable the **Full Preview** checkbox.

Click **OK**.

Rotate / zoom to different angles to verify the fillets. Ensure that they were added to the 7 edges as noted.

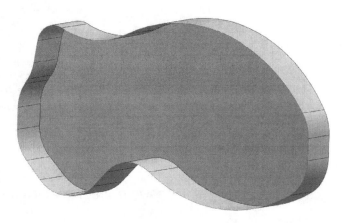

6. Adding another .500" fillets:

Select the **Fillet** command once again.

Use the default **Constant Size** radius option once again.

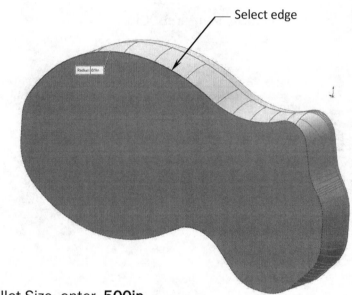

— Select edge

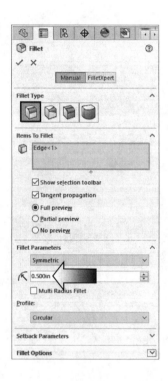

For Fillet Size, enter **.500in**.

For Items to Fillet, select the **edge** as noted.

The preview graphics shows the fillet is being propagated along the perimeter of the surface model.

Click **OK**.

Verify that the fillet is added to all edges along the back side of the surface model.

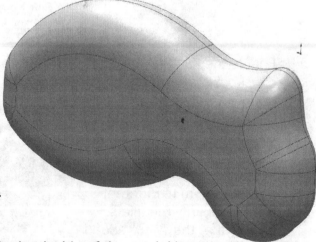

(Note: the fillet is shown from the back side of the model.)

7. Thickening the surface model:

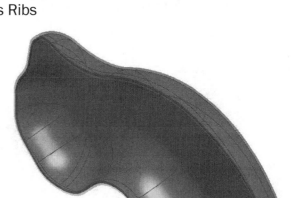

After completing steps 1 through 6 the surface model can now be thickened, allowing other solid features such as Ribs and Mounting Bosses to be added.

Select the **Thicken** command from the **Surfaces** tab.

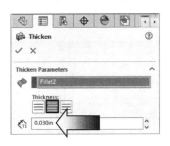

For Thicken Parameters, select the **Surface Model** either from the graphics area or from the FeatureManager tree.

For Thickness location, select the option for **Both Sides**.

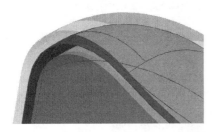

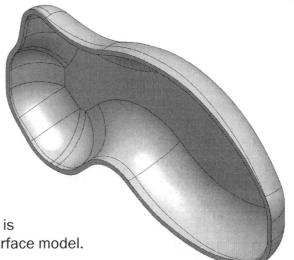

For Thickness, enter **.030in**. (The total thickness is .060in.)

Zoom in to verify that the thickness is being added to <u>both sides</u> of the surface model.

Click **OK**.

8. Adding the inner support feature:

Select the Top plane and open a **new sketch**.

Sketch the profile and add the dimensions / relations shown below.

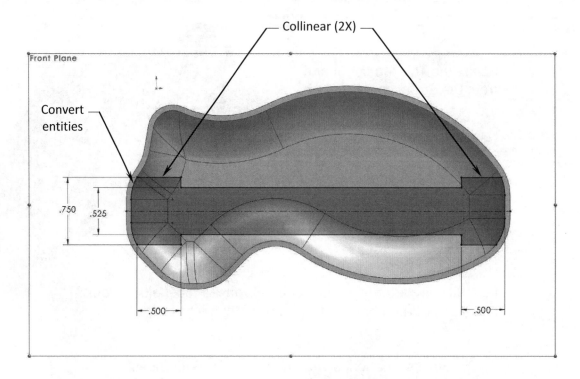

Switch to the **Features** tab and click the **Extruded Boss-Base** command.

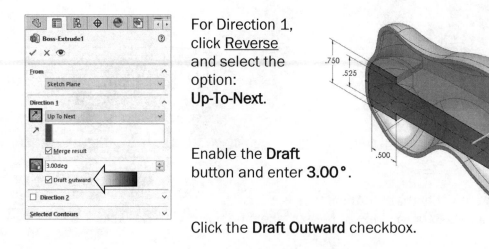

For Direction 1, click Reverse and select the option: **Up-To-Next**.

Enable the **Draft** button and enter **3.00°**.

Click the **Draft Outward** checkbox.

Click **OK**.

9. Creating an extruded cut:

Select the <u>Right</u> plane and open a **new sketch**.

Sketch a **Circle** as shown.

Add a **Vertical relation** between the **center** of the circle and the **Origin**.

Add the size and location dimensions to fully define the sketch.

Switch to the **Features** tab and click **Extruded Cut**.

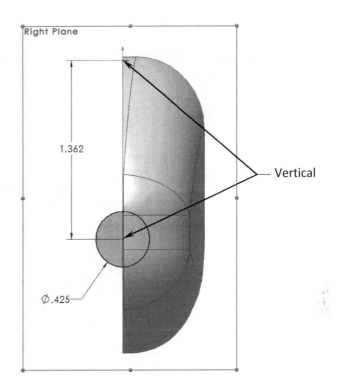

For Direction 1, select **Through All –Both**.

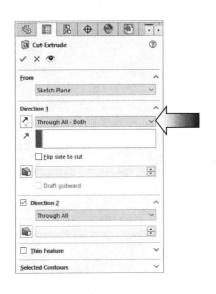

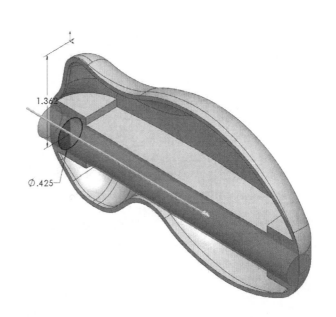

Click **OK**.

10. Adding the support ribs:

Open a **new sketch** on the <u>Front</u> plane.

Sketch the **3 rib profiles** shown below. Converting the inner or outer curves and using trim can help speed up the sketching process.

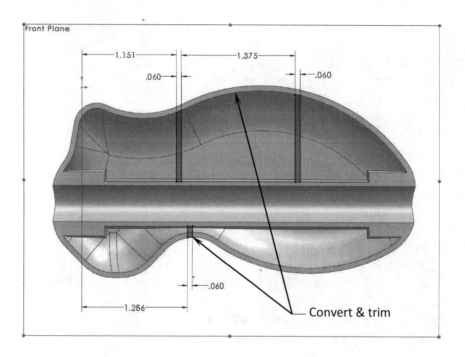

Front Plane

1.151 .060 1.375 .060

1.286 .060

— Convert & trim

Switch to the **Features** tab and click **Extruded Boss-Base**.

For Direction 1, click <u>Reverse</u> and select: **Up-To-Next**.

Enable the Draft button and enter **3.00deg**.

Click **Draft Outward**.

Click **OK**.

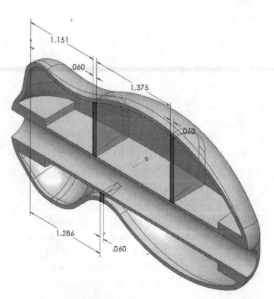

1.151 .060 1.375 .060 1.286 .060

11. Adding the mounting bosses:

Select the <u>Front</u> plane and open a **new sketch**.

Sketch **4 circles** and add the dimensions / relations shown below.

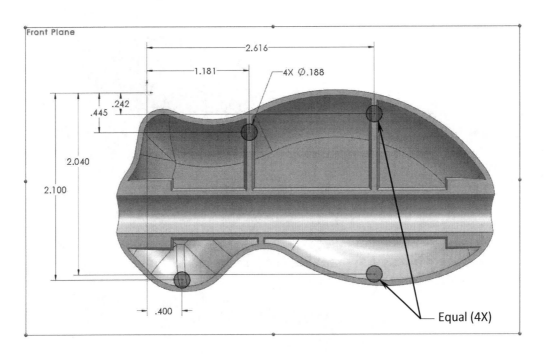

Switch to the **Features** tab and click the **Extruded Boss-Base** command.

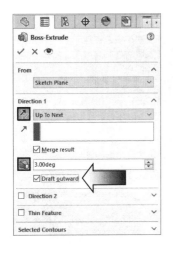

For Direction 1, click <u>Reverse</u> and select: **Up-To-Next**.

Enable the Draft button and enter: **3.00deg**.

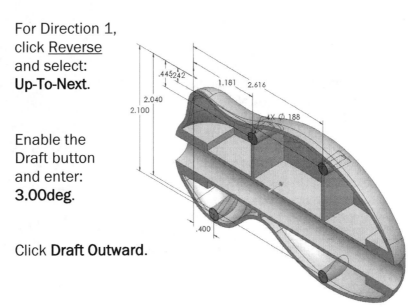

Click **Draft Outward**.

Click **OK**.

12. Adding fillets:

Due to the complex blends at the corners, it is best to create two separate fillets around the ribs and mounting bosses rather than making them in the same fillet.

Select the **Fillet** command.

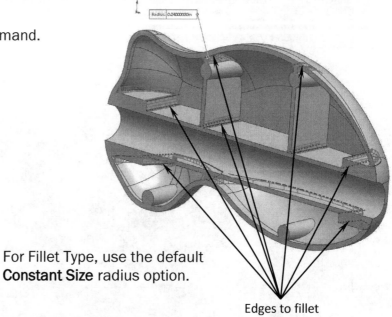

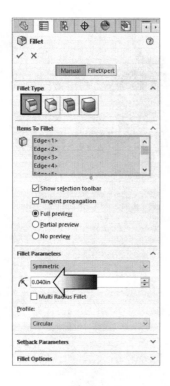

For Fillet Type, use the default **Constant Size** radius option.

For Fillet size, enter **.040in**.

For Items to Fillet, select the **edges** shown above.

Click **OK**.

Edges to fillet

Once again, select the **Fillet** command and apply the same fillet to the remaining edges of the Ribs and Mounting Bosses.

Zoom / Rotate and view the model from different angles to verify that all edges have the same fillet applied to them.

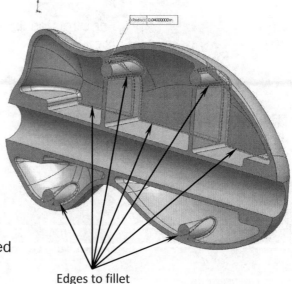

Edges to fillet

13. Saving the housing right-half:

Select **File, Save As** and enter: **Handle Housing_Right** for the name of the file.

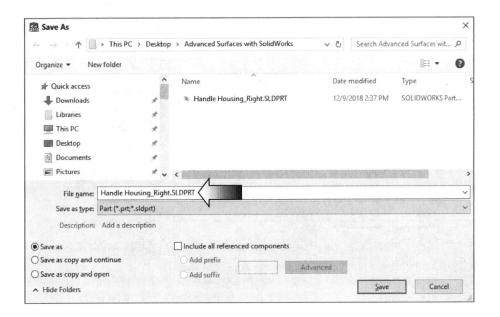

14. Mirroring the part:

Select the **Front** plane and click: **Insert, Mirror Part** (arrow).

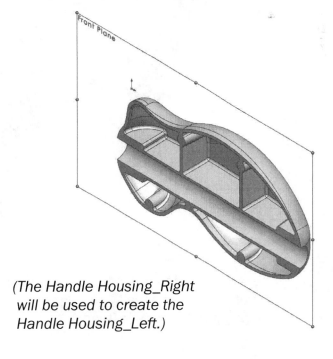

(The Handle Housing_Right will be used to create the Handle Housing_Left.)

In the Insert Part dialog, enable the **Break Link to Original Part** checkbox.

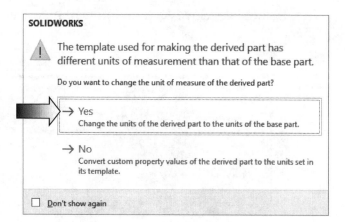

Click **YES** to allow the use of the same Template and Units to the Mirrored part (arrow).

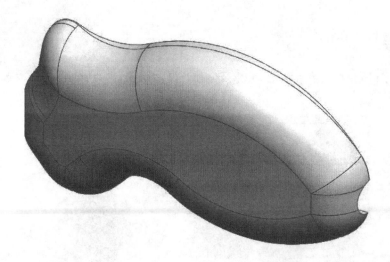

The mirrored part is created and opened in its own window.

It is necessary to enable the **Break Link to Original Part** option because the mirrored part will have <u>Mounting Holes</u> added while the other part will receive the <u>Mounting Bosses</u> instead.

15. Creating an extruded cut feature:

Open a **new sketch** on the <u>Front</u> plane.

Select the **Polygon** command (arrow) and use the default **6-sided** polygon option (arrow).

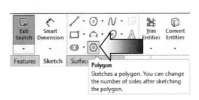

Sketch **centerlines** to assist with centering the polygons.

Sketch **8 Polygons** as shown in the image below.

Add a **Horizontal relation** to one line in each polygon.

Add the dimensions / relations indicated to fully define the sketch.

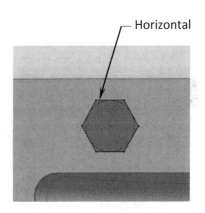

Horizontal

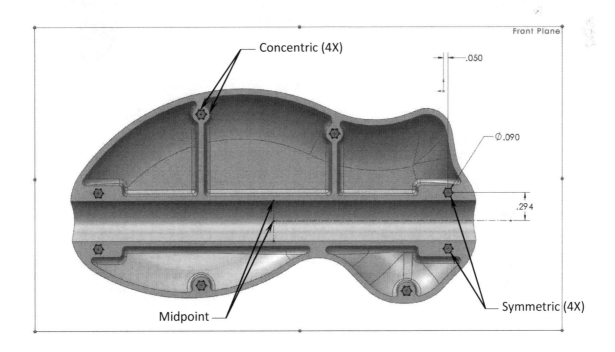

Front Plane

Concentric (4X)

.050

Ø.090

.294

Midpoint

Symmetric (4X)

Switch to the **Features** tab and click **Extruded Cut**.

For Direction 1, use the default **Blind** command.

For Depth, enter: **.190in**.

Enable the Draft button and enter: **1.00deg**.

Click **OK**.

The Handle Housing_Left is now completed, and later, after the mounting pins are added to the part, it will be assembled with the Handle Housing_Right,

16. Saving the left half:

Click **File, Save**.

Enter **Handle Housing_ Left** for the file name.

Close the current document. Next, we will focus on adding mounting pins to the Handle Housing_Right.

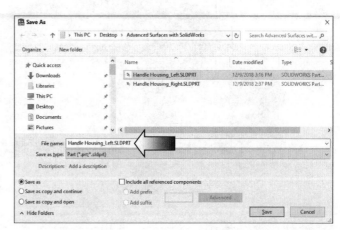

Adding the alignment pins:

Push **Control+Tab** to switch to the **Handle Housing_Right** document.

Select the <u>Front</u> plane and open a **new sketch**.

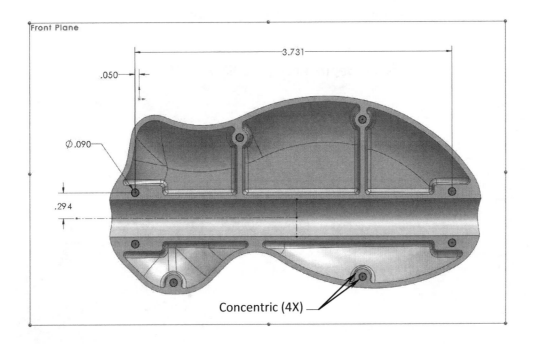

Sketch **8 Circles** and add the same dimensions / relations as shown in the previous step.

Switch to the **Features** tab.

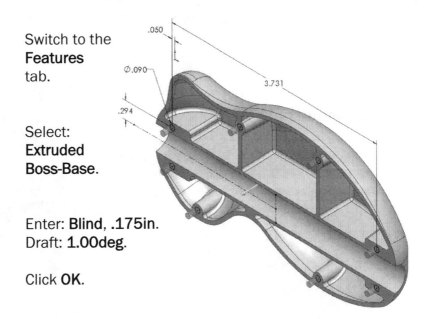

Select: **Extruded Boss-Base.**

Enter: **Blind, .175in.**
Draft: **1.00deg**.

Click **OK**.

17. Adding chamfers:

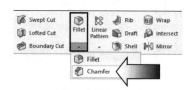

Select the **Chamfer** command from the Fillet drop-down menu (arrow).

For Chamfer Type, use the default **Angle and Depth** option.

For Chamfer Angle and Depth, enter: **.020in**. and **45deg**.

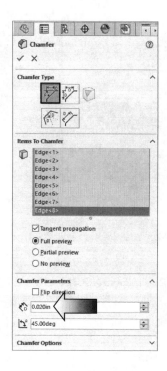
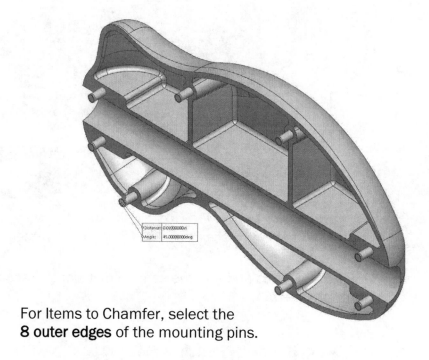

For Items to Chamfer, select the **8 outer edges** of the mounting pins.

Click **OK**.

18. Saving the right half:

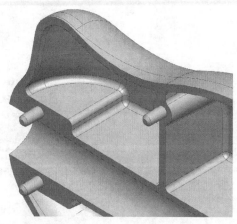

Click **File, Save** (Handle Housing_Right).

Keep the same file name and overwrite the previous document with this latest one.

Next, we will insert the Handle Housing_Left into this document.

19. Inserting a part document:

Click **Insert, Part** (arrow).

Select the **Handle Housing_Left** document and click **Open**.

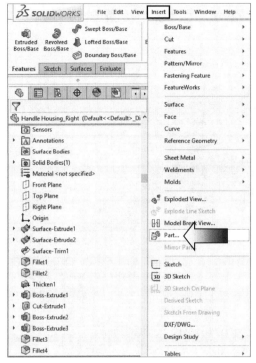

In the Insert Part dialog <u>enable</u> the options as shown.

Click the <u>green check mark</u> to place the part on the same **origin** as the other part.

Turn off the **View Planes** option, on the View Heads-Up, to hide the planes (arrow).

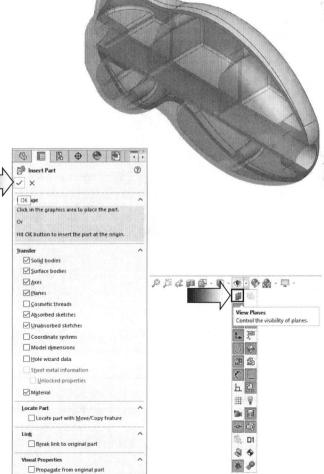

20. Creating an exploded view:

Select **Insert, Exploded View** (arrow).

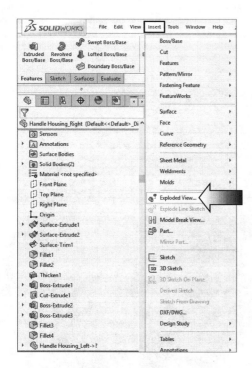

Select the **Handle Housing_Left** from the graphics area.

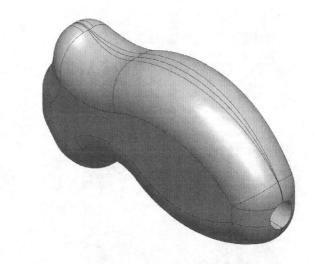

Drag the **Z arrowhead** toward
the left side, approximately **2.40in.** or **2.50in.**

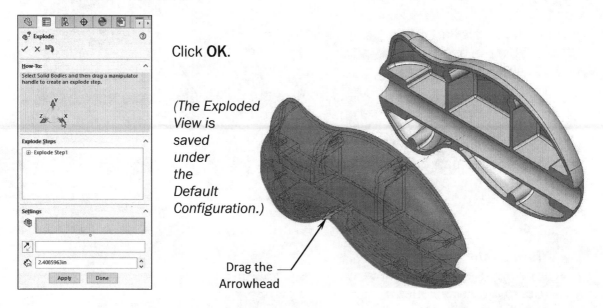

Click **OK.**

*(The Exploded
View is
saved
under
the
Default
Configuration.)*

Drag the ——
Arrowhead

Re-save and close all documents.

Exercise: Hybrid Modeling

The exercises give you an opportunity to apply what you have learned from the lessons. The instructions are limited to allow you to try things out using your own techniques, at your own pace.

1. Opening a part document:

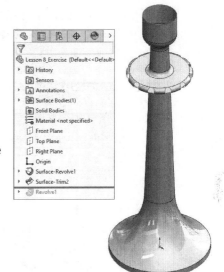

Browse to the Training Folder and open a part document named: **Lesson 9_Exercise.sldprt.**

There is 1 surfaces body and 1 solid body in this model. This exercise will teach us a unique method to combine the use of surfacing and solid modeling to complete the design.

2. Creating a Filled Surface:

There are 12 openings around the flange. We will fill one of the openings and then circular pattern it 12 times around the part.

Switch to the **Surfaces** tab.

Click **Filled Surface.**

For Patch Boundary, select **2 edges** of an opening on the flange as noted.

For Curvature Control, use the default **Contact** and set other options as shown in the dialog.

Click **OK.**

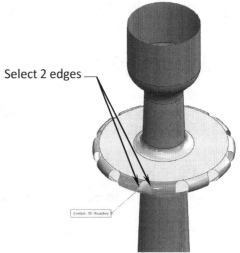

Select 2 edges

3. Creating a Circular Pattern:

Switch to the **Features** tab.

Click **Circular Pattern** under the Linear Pattern drop-down list.

For Pattern Direction, select the <u>circular edge</u> as noted.

Enable **Equal Spacing**.

For Number of Instances, enter **12**.

For Bodies to Pattern, select the **Surface-Fill1** either from the FeatureManager tree or from the graphics area.

Click **OK**.

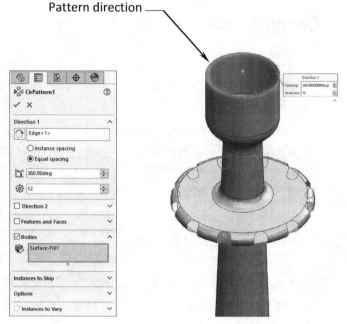

Pattern direction

4. Knitting the surfaces:

Switch back to the **Surfaces** tab.

Click **Knit Surface**.

For Selections, select the **main body** and all other patterned surfaces. <u>13 surfaces</u> all together.

Enable the **Merge Entities** checkbox.

Click **OK**.

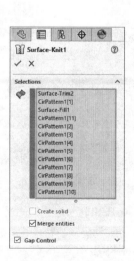

5. Adding fillets:

Click **Fillet**.

Use the default **Constant Size Fillet** option.

For Items to Fillets, select all edges of the 12 patches.

For Radius, enter **.125in**.

Click **OK**.

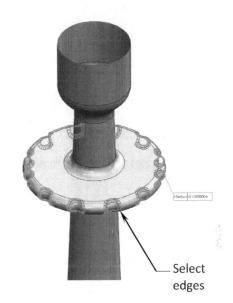

Select edges

6. Thickening the surface body:

Click **Thicken**.

For Thicken Parameters, select the the **surface body** from the graphics area.

For Thickness Direction, select **Inside**.

For Wall Thickness, enter **.100in**.

Click **OK**.

The surface body is thickened to a solid body.

7. Showing a solid body:

The candle spike was created as a solid feature, and it was in the rolled back state.

Drag the Rollback Line down <u>below</u> the **Revolve1** feature.

The Solid Bodies folder on the FeatureManager tree shows 2 solid bodies exist in the model.

We will combine them into a single body in the next step.

8. Combining the solid bodies:

Select **Insert, Features, Combine.**

For Orientation Type, click **Add.**

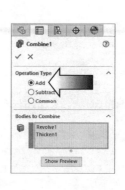

For Bodies to Combine, select the **Main Body** and the **Revolved Body.**

Click **OK.**

The 2 solid bodies are combined into a single body.

9. Saving your work:

Select **File, Save As**.

Enter **Lesson 9_Exercise_Completed** for the file name.

Click **Save**.

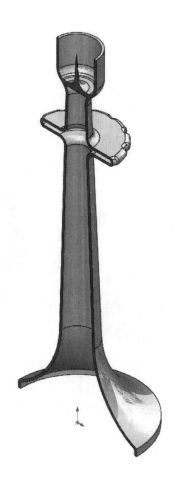

Exercise: Using the Wrap tool

The Wrap tool wraps a sketch onto a planar or non-planar face. The options available with the Wrap tool are Emboss, Deboss, and Scribe.

1. Opening a part document:

Browse to the Training Folder and open a part document named: **Wrap.sldprt**

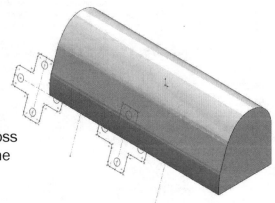

The model contains 2 sketches; one of them will be used to create a Wrap-Emboss feature, and the other is used to make the Wrap-Deboss feature.

2. Creating a Wrap-Emboss feature:

Select the sketch named: **Wrap Boss** from the FeatureManager tree and click: **Insert, Feature, Wrap**.

For Wrap Type, click: **Emboss**.

For Wrap Method select: **Spline Surface**.

For **Face For Wrap Sketch** select the face as noted.

Enter: **.100in** for Thickness.

For Pull Direction, select **Plane1** from the FeatureManager tree.

Click **OK**.

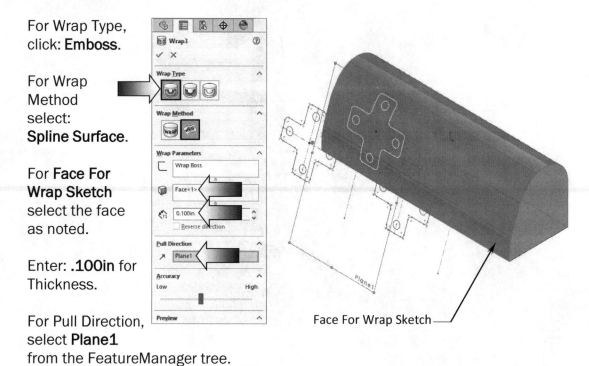

Face For Wrap Sketch

3. Creating a Wrap-Deboss feature:

Select the sketch named: **Wrap Cut** from the FeatureManager tree and click:
Insert, Feature, Wrap.

For Wrap Type,
click: **Deboss.**
For Wrap Method
select:
Spline Surface.

For **Face For
Wrap Sketch**
select the <u>front
face</u> as noted.

Enter: **.100in** for
Thickness.

For Pull Direction,
select **Plane1** from the
FeatureManager tree.

Click **OK.**

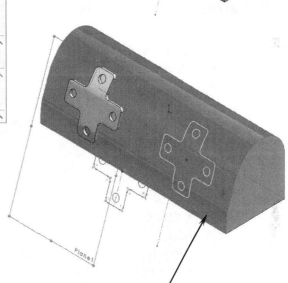

Face For Wrap Sketch

Compare your model with the images
shown on the lower right side and
make any adjustments if needed.

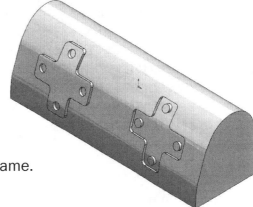

4. Saving your work:

Select **File, Save As.**

Enter: **Wrap_Completed.sldprt** for the file name.
Click **Save.**

Exercise: Using Offset From Surface

This exercise will teach us one of the easier methods to extrude the text and match the curvature of a non-planar surface. All letters in the sketch are considered 1 entity and they will have the same thickness when extruded.

1. Opening a part document:

Browse to the Training Folder and open a part document named: **Text On Curved Surfaces.sldprt.**

2. Extruding the 1st text:

Select the sketch named: Solidworks from the FeatureManager tree and click: **Extruded Boss/Base.**

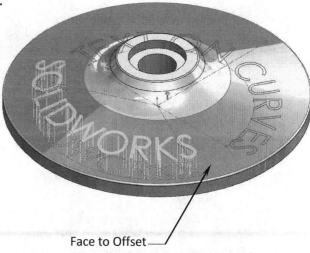

For Direction 1 select: **Offset From Surface.**

For **Face to Offset**, select the <u>upper surface</u> of the model as noted.

For Offset Distance, enter **.050in.**

Enable the checkboxes: **Reverse Offset** and **Translate Surface.**

Face to Offset

Click **OK.**
(Hide the sketch Solidworks for clarity.)

3. Extruding the 2ⁿᵈ text:

Select the sketch named **Text On Curves** from the FeatureManager tree.

Click **Extruded Boss/Base**.

For Direction 1 select: **Offset From Surface** from the drop-down list.

For **Face to Offset**, select the <u>upper surface</u> of the model as noted.

For Offset Distance, enter **.050in**.

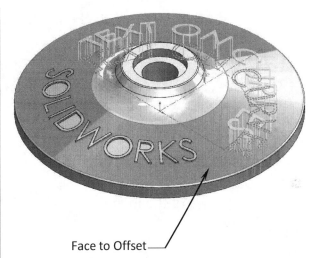

Face to Offset

Enable the checkboxes: **Reverse Offset** and **Translate Surface**.

Click **OK**.

Inspect your model against the image shown here.

4. Saving your work:

Select **File, Save As**.

Enter: **Text On Curved Surfaces_Completed.sldprt** for the file name.
Click **Save**.

1. Opening a part document:

Browse to the Training Folder and open a part document named: **Offset From Surface.sldprt**.

2. Extruding the text:

Select **Sketch2** from the FeatureManager tree and click: **Extruded Boss/Base**.

For Direction 1 select: **Offset From Surface** from the drop-down list.

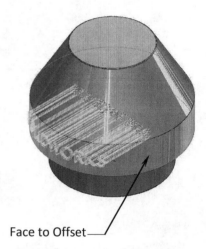

For **Face to Offset**, select the <u>side face</u> of the model as indicated.

For Offset Distance, enter: **.025in**.

Enable the checkboxes: **Reverse Offset** and **Translate Surface**.

Face to Offset

Click **OK**.

The sketch-text is protruded outward and also matches the curvature of the selected face.

3. Saving your work:

Select **File, Save As**.

Enter: **Offset From Surface_Completed.sldprt** for the file name.

Click **Save**.

Chapter 10: Mold Tools, Intersect & Core/Cavity
Molded Parts

This lesson has two sections. In the first section we will create a plastic part using surface tools. The second section will use the same part to create the core and cavity mold.

1. Starting a new part document:

Select the Front plane and open a **new sketch**.

Sketch the outline of the main sketch using the **Line** and **3-Point Arc** commands.

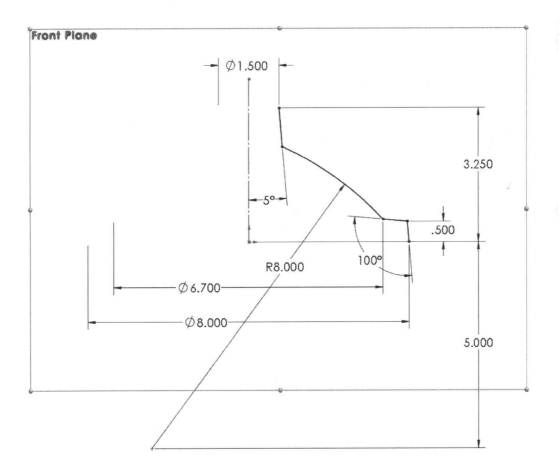

Add the dimensions shown in the image to fully define the sketch.

2. Creating a revolved surface:

Switch to the **Surfaces** tab.

Click **Revolved Surface**.

Use the default **Blind** type.

Also use the default **360°** revolve angle.

Click **OK**.

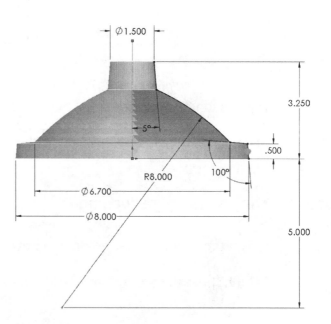

3. Making the sketch of the sweep path:

Open a **new sketch** on the <u>Front</u> plane.

Sketch a **3-Point Arc** and add dimensions to fully define the sketch.
(The dimension R8.250 is a reference dimension; it is shown for clarity only.)

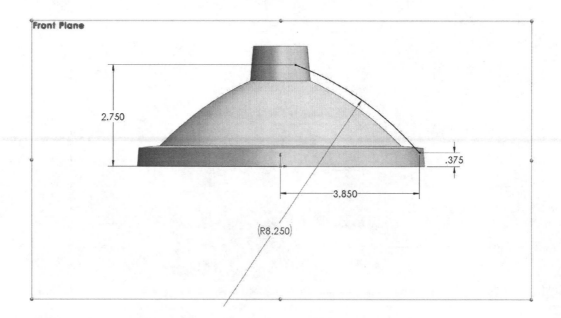

<u>Exit</u> the sketch.

4. Creating a new plane:

Select the **Plane** command from the **Reference Geometry** drop-down list.

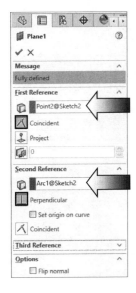

For First Reference,
select the **3-Point Arc**.

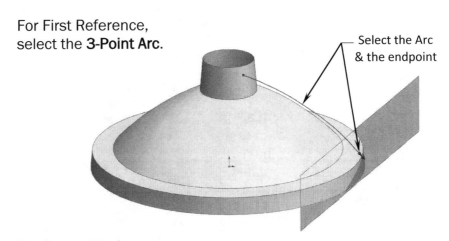

Select the Arc
& the endpoint

For Second Reference,
select the **end point** on the right end of the arc.

A new plane which is Coincident to the end point and Perpendicular to the arc is being created.

Click **OK**.

Select the <u>new plane</u> and open a **new sketch**.

Sketch the profile shown below and add the dimensions / relations to fully define the sketch.

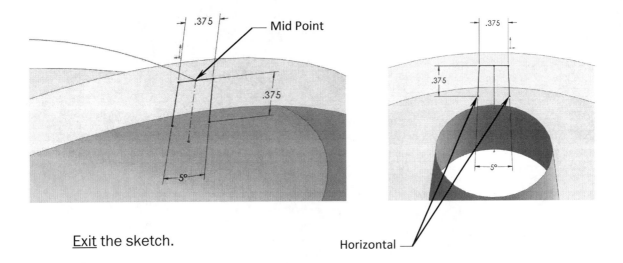

<u>Exit</u> the sketch.

5. Creating a rib feature:

Switch back to the **Surfaces** tab and click **Swept Surface**.

For Sweep Profile, select **Sketch3** (the U-shaped sketch).

For Sweep Path, select the **3-Point Arc**.

Keep all other parameters at their default values.

Click **OK**.

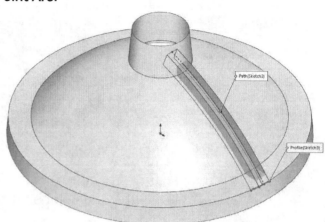

6. Circular patterning the rib:

Change to the **Features** tab and select the **Circular Pattern** command from the Linear Pattern drop-down menu.

For Direction 1, select the **circular edge** as noted.

Select / enable the following:

* **Equal Spacing**
* **360deg.**
* **5 instances**

Enable the **Bodies** checkbox and select the **Rib** feature.

Click **OK**.

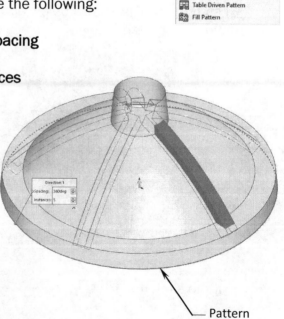

Pattern Direction

7. Trimming the overlaps:

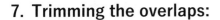

The ribs are overlapping and intersecting the main surface body. This will cause problems when adding a wall thickness to the model. They will need to be trimmed.

Switch to the **Surfaces** tab and click **Trim Surface**.

For Trim Type, click **Mutual**.

For Trimming Surfaces, select the **Main body** and the **5 Ribs**.

Select the main surface body and the 5 ribs

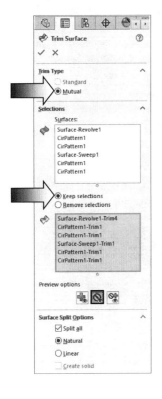

For Keep Selections, select the **Main body** and the **top 5 surfaces** of the ribs.

Select the main surface body and the top 5 surfaces of the ribs

Keep all other parameters at their default values.

Click **OK**.

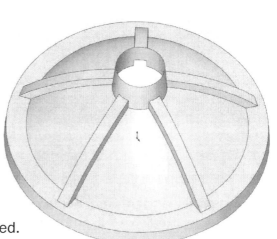

Zoom in closer and check the model to ensure that the overlaps are removed.

8. Adding thickness:

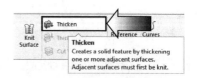

A wall thickness can be added to the surface model at this point so that the core and cavity mold can be made from it.

Click **Thicken** on the **Surfaces** tab.

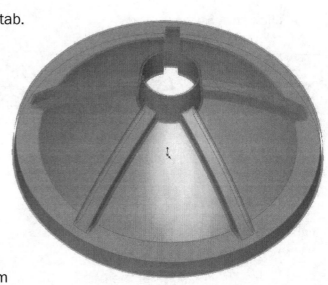

For Thicken Parameter, select the **Surface model** directly from the graphics area.

For Thickness Location, select **Both Sides**.

For Wall Thickness, enter: **.060in** (.120in. total).

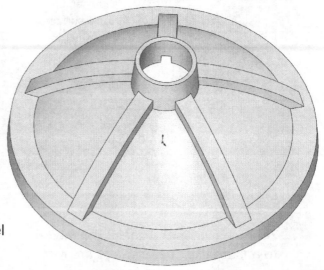

Click **OK**.

A wall thickness of .120in is added and the surface model is converted to a solid model.

9. Adding the .060" fillets:

Fillets and rounds serve different purposes throughout the designs. The fillets we are adding to the model at this point are meant to break the sharp edges and the cavity mold can be filled more easily during the mold-process.

Click **Fillet**.

Use the default **Constant Size Radius** option.

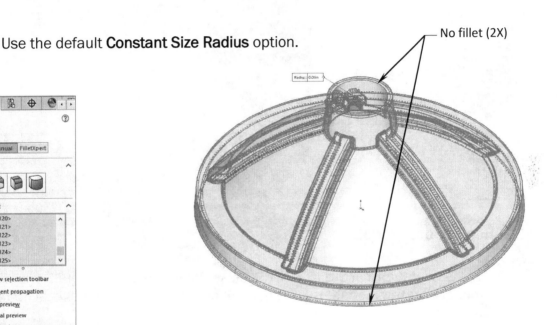

No fillet (2X)

For Radius size, enter **.060in**.

For Items to Fillet, select **all edges** in the model (Control+A), <u>except</u> for the <u>2 edges</u> as noted.

Click **OK**.

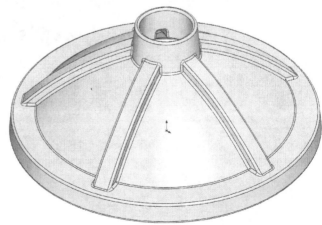

Again, zoom in closer and check the model to ensure all edges, except for the two mentioned in the note, are filleted.

10. Adding material:

Right-click the **Material** option on the FeatureManager tree and select:
Edit Material.

Under **SOLIDWORKS Materials**, expand the **Plastic** folder and select: **ABS**

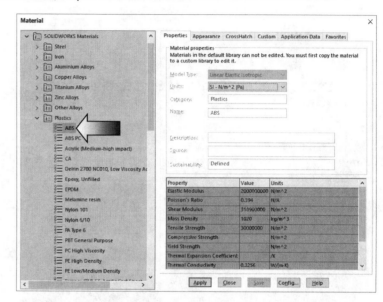

11. Scaling the model:

The Scale feature in SOLIDWORKS is used to increase the overall size of the model to accommodate the mold shrinkage. The actual dimensions in the model will not change.

Right-click the Surfaces tab and enable the **Mold Tools**.

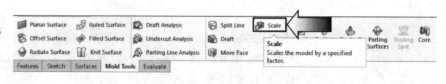

Switch to the **Mold Tools** tab and click **Scale**.

Use the default **Centroid** for Scale About.

Enter: **1.015** for Scale (1.5% larger).

Click **OK**.

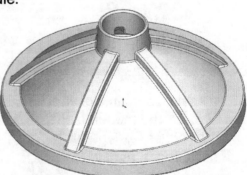

12. Creating the parting lines:

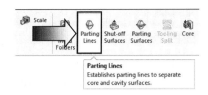

Parting Lines
Establishes parting lines to separate core and cavity surfaces.

Select the **Parting Lines** command on the **Mold Tools** tab.

For Direction of Pull, select the **Top** plane from the FeatureManager tree.

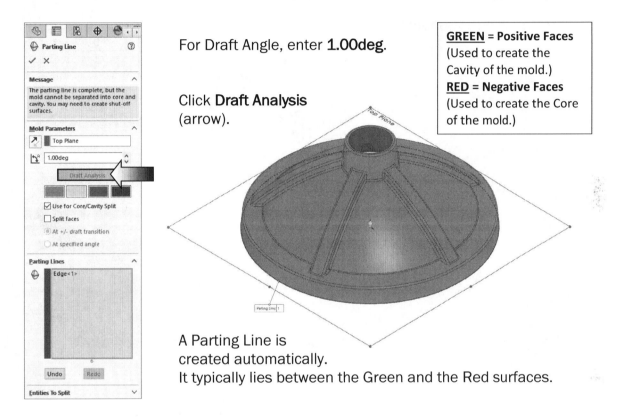

For Draft Angle, enter **1.00deg**.

Click **Draft Analysis** (arrow).

> **GREEN** = Positive Faces
> (Used to create the Cavity of the mold.)
> **RED** = Negative Faces
> (Used to create the Core of the mold.)

A Parting Line is created automatically.
It typically lies between the Green and the Red surfaces.

In a more complex mold, the parting lines can be created manually and additional surfaces can be added to ensure the proper shut-off between the core and cavity.

Click **OK**.

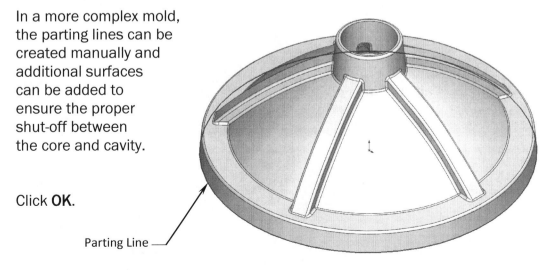

Parting Line

13. Creating the shut-off surfaces:

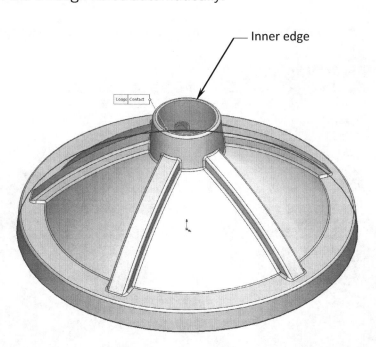

There should be no through holes in a
molded part. The Shut-off surfaces command
was designed to close-off the through holes automatically.

Inner edge

Click **Shut-Off Surfaces**.

The through hole (on top) was found and shut-off with a
surface.

<u>Enable all checkboxes</u> for Knit, Filter Loops, Show Preview, and Show Callouts.

Only one Shut-off Surface
feature is allowed in a model.
For fill type, select:
Contact to fill the
through hole.

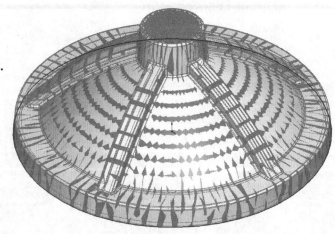

Click **OK**.

14. Creating the parting surfaces:

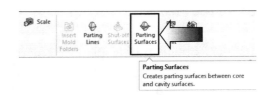

The Parting surfaces split the mold cavity from the core.

Parting Surfaces
Creates parting surfaces between core and cavity surfaces.

Click **Parting Surfaces** on the **Mold Tools** tab.

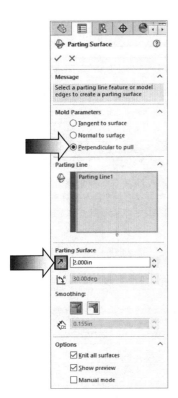

For Mold Parameters, select: **Perpendicular to Pull.**

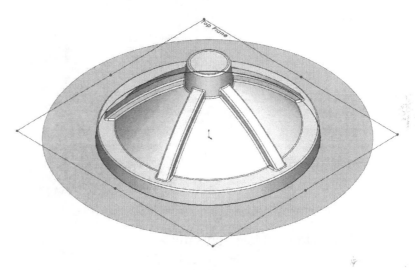

The **Parting Line1** should be selected automatically.

For Parting Surface Distance, enter: **2.000in**, and click **Reverse** (arrow).

For Smoothing, use the default **Sharp** option.

Enable the **Knit All Surfaces** checkbox.

Click **OK.**

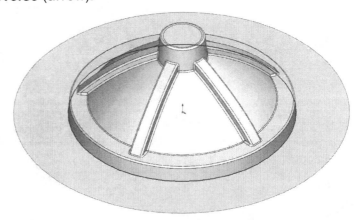

15. Creating a tooling split:

A tooling split is inserted to create the core and cavity blocks for a mold. To create a tooling split, the part must have at least three surface bodies in the Surface Bodies folder: a Core surface body, a Cavity surface body, and a Parting Surface body.

Select the <u>Parting Surface</u> as the sketch plane and open a **new sketch.**

Sketch a **Circle** and add the **Ø12.00in.** diameter dimension.

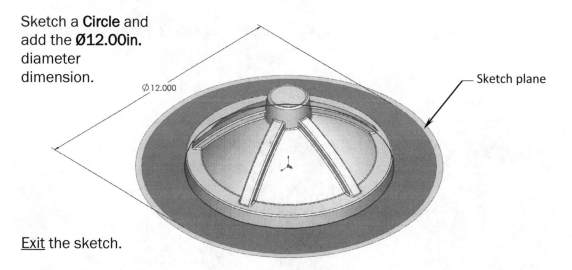

<u>Exit</u> the sketch.

Switch to the **Mold Tools** tab and click **Tooling Split.**

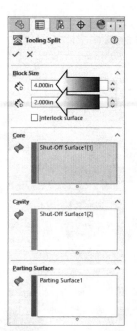

For Block Size, enter:
    4.000in for upper block
    2.000in for lower block

Click **OK.**

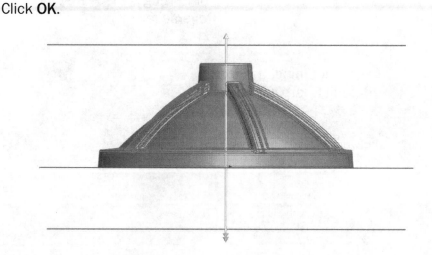

16. Creating an exploded view:

An exploded view is created to verify the internal details of the Core and Cavity.
(Note: While the exploded view is active, all toolbars are temporarily disabled.)

Exploded views are stored on the Configuration-Manager, under the Default configuration.

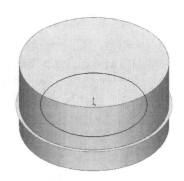

Select **Insert, Exploded View** (arrow).

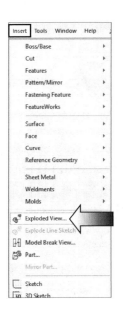

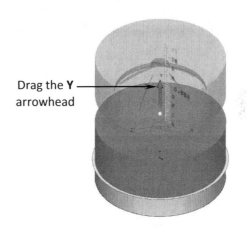

Drag the **Y** arrowhead

Select the **upper block** and drag the **Y arrowhead** upward, between 9 and 10 inches. This creates the Step1 in the Explode Steps dialog. Click **Done**.

Select the **lower block** and drag the **Y arrowhead** downward, also between 9 to 10 inches. The Step2 is created and stored on the Explode Steps dialog. Click **Done**.

Click **OK** to exit the Exploded View mode.

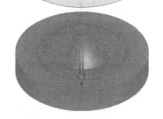

17. Renaming and hiding the references:

Reference surfaces such as Shut-Off Surfaces and Parting surfaces are created to assist the mold creation processes. They are stored in the Surface Bodies Folder and hidden when the mold is completed.

Click the **Surface Bodies** folder on the FeatureManager tree and select **Hide**.

Click the **Parting Surface** in the graphics area and select **Hide** (arrow).

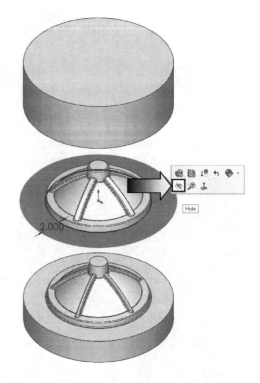

Expand the **Solid Bodies** folder and rename the 3 solid bodies as follows: **Part, Cavity**, and **Core**.

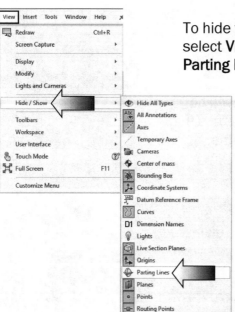

To hide the Parting Lines, select **View, Hide / Show, Parting Lines** (arrow).

18. Saving your work:

Save your document as:
Molded Part.sldprt.

Close all documents.

Using Intersect

Intersect is an alternative option to the Trim command. It offers several operations within the command such as Split, Trim, Knit, and thicken the surface bodies into a single solid body.

1. Opening a part document:

Open a part document named: **Core.sldprt**

This model was derived from the previous lesson to show how a part can be made from a hollow, empty space between the two solid bodies: Core and Cavity.

2. Inserting another part document:

Select **Insert, Part** (arrow).
Locate and open the part document named: **Cavity.sldprt**

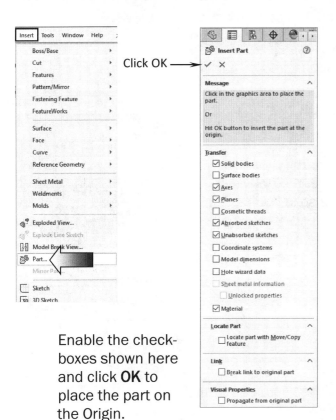

Enable the check-boxes shown here and click **OK** to place the part on the Origin.

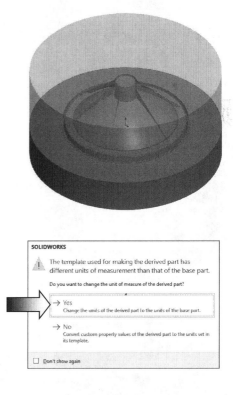

Click **YES** to close the dialog box.

3. Moving the solid bodies:

The command Move/Copy Body is used as an alternative option to view the interior details of the 2 solid bodies.

Select **Insert, Features, Move/Copy**.

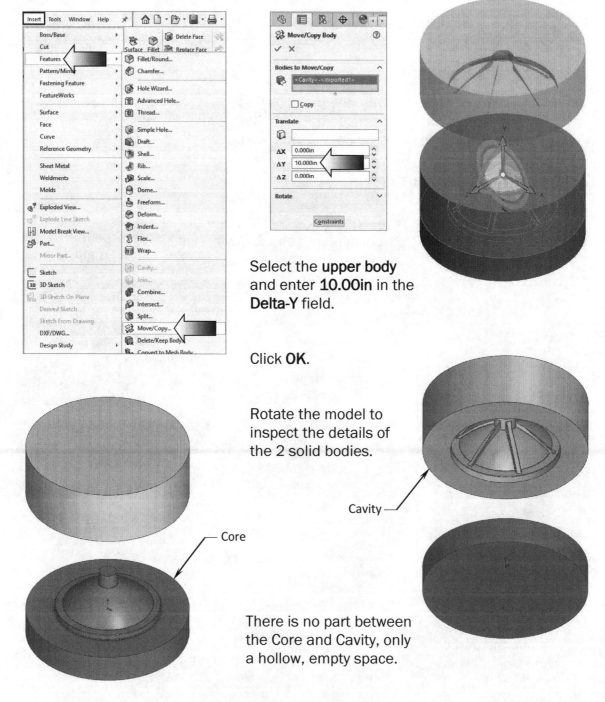

Select the **upper body** and enter **10.00in** in the **Delta-Y** field.

Click **OK**.

Rotate the model to inspect the details of the 2 solid bodies.

Cavity —

— Core

There is no part between the Core and Cavity, only a hollow, empty space.

4. Using the Intersect option:

The Intersect command combines the Trim, Knit, and Thicken options into a single operation. In this case study, it can convert the empty space between the two solid bodies into a new, single body while removing the others.

<u>Delete</u> the **Body-Move/Copy** feature from the Feature tree; it is no longer needed.

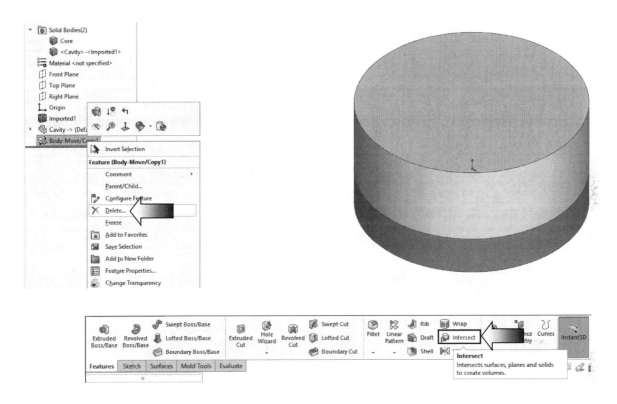

Switch to the **Features** tab and click **Intersect** (arrow).

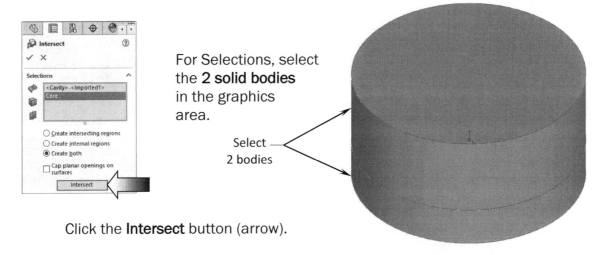

For Selections, select the **2 solid bodies** in the graphics area.

Select — 2 bodies

Click the **Intersect** button (arrow).

For Regions to Exclude, select **Region 1** (the upper block) and **Region 3** (the lower block).

The preview graphics show the 2 upper and lower blocks are being removed.

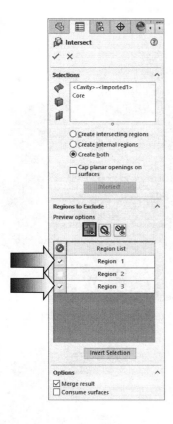

Region 2 is the new solid body being created from the hollow space between the upper and lower blocks.

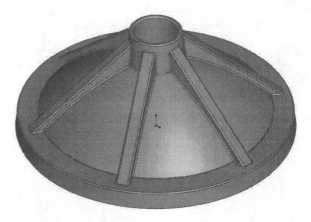

Under Options, <u>enable</u> the **Merge Result** checkbox.

<u>Clear</u> the **Consume Surfaces** checkbox.

Click **OK**.

A new solid body is created.

5. **Saving your work:**

Select **File, Save As**.

Enter: **Intersect_Completed** for the file name.

Click **Save**.

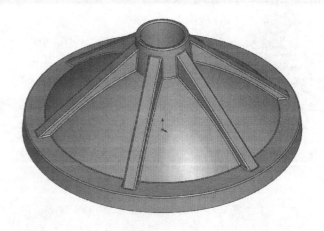

Exercise: Core & Cavity Creation

The exercises give you an opportunity to apply what you have learned from the lessons. The instructions are limited to allow you to try things out using your own techniques, at your own pace.

1. Opening a part document:

Browse to the Training Folder and open a part document named: **Lesson 10_Exercise.sldprt.**

There is only 1 surface in this model. We will add the final touches and then create a core and cavity mold for it.

2. Adding drafts:

Switch to the **Features** tab and click **Draft.**

For Type of Draft, use the default **Neutral Plane** option.

For Draft Angle, enter **3°.**

For Neutral Plane select the **Top Plane.**

For Faces to Draft, select the **2 side faces.**

Click **OK.**

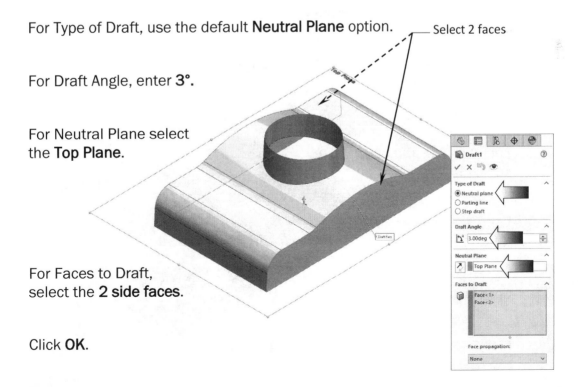

3. Adding fillets:

Click **Fillet**.

For Fillet Type use the default **Constant Size**.

For Items to Fillet, select the _face_ as indicated.

For Radius, enter **.250in**

Click **OK**.

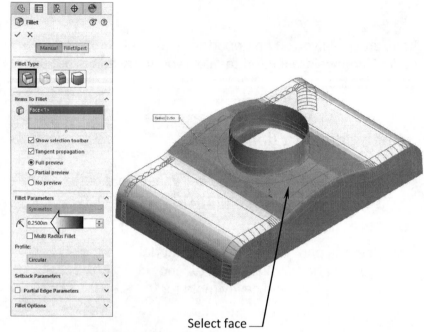

Select face ⌐

4. Thickening the surface model:

Click **Thicken** on the **Surfaces** tab.

For Thicken Parameters _select the model_ from the graphics area.

For Thickness Direction select the _Inside_ button.

For Wall Thickness, enter: **.080in**.

Click **OK**.

Ensure the thickness is added to the _inside_.

5. Scaling the model:

The scale feature is used to accommodate the shrink rate of the material, it does not change the actual dimensions of the part.

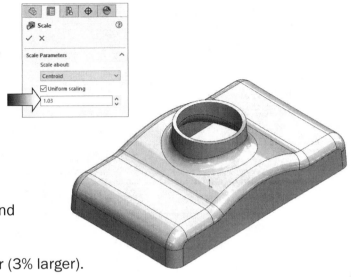

Switch to the **Mold Tools** tab and click **Scale**.

Use the default **Centroid** and **Uniform Scaling**.

Enter **1.03** for Scale Factor (3% larger).

Click **OK**.

6. Creating the parting lines:

Click **Parting Lines**.

For Pull Direction select the **Top** plane.

For Draft Angle, enter **2°**.

Click the **Draft Analysis** button.

The Parting is created automatically. It runs along the bottom edges of the part.

Enable the checkbox **Use for Core/Cavity Split**.

Click **OK**.

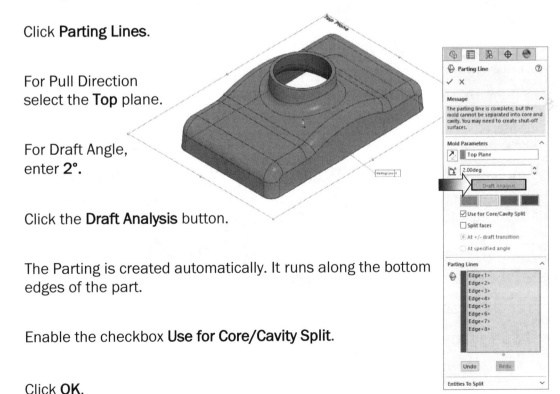

7. Shutting off the through hole:

Shut-Off surfaces are only required for through holes.

Click **Shut-Off Surfaces**.

The command shuts the opening of the hole with a surface.

Enable the check-boxes shown.

Click **OK**.

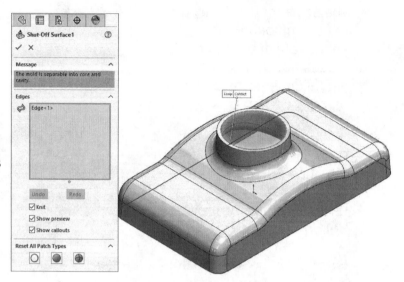

8. Creating the parting surface:

Click **Parting Surfaces**.

Enter/select the following:

Perpendicular to Pull.

Parting Surface Distance: **2.00in**

Smoothing: **Sharp**.

Knit all Surfaces.

Show Preview.

Click **OK**.

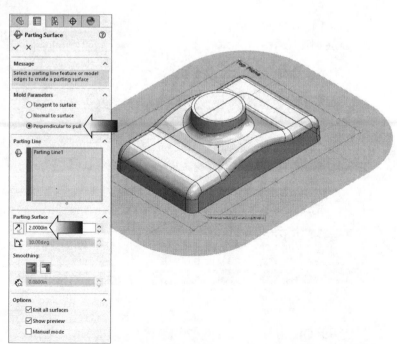

9. Making the mold block sketch:

Open a **new sketch** on the <u>Parting Surfaces</u>.

Sketch a **Rectangle**. <u>Avoid</u> snapping the corners of the rectangle to any edges of the Parting Surfaces.

Add the dimensions shown to fully define the sketch.

<u>Exit</u> the sketch but keep it selected or highlighted.

10. Creating the tooling split:

Switch to the **Mold Tools** tab.

Click **Tooling Split**.

For Block Size enter:

2.50 (upper block)

1.50 (lower block)

Click **OK**.

The Core and Cavity mold blocks are created but in the closed position.

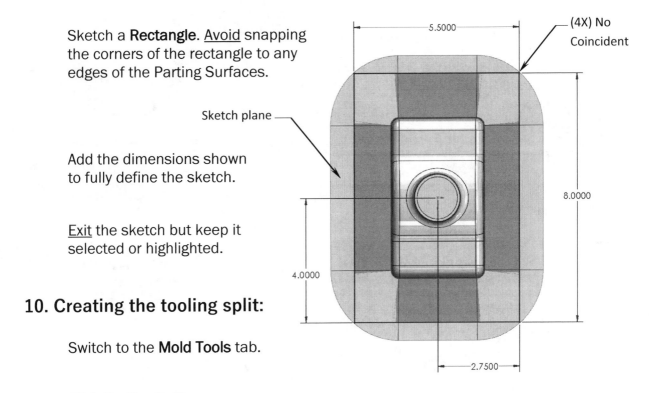

We need to separate them to see the interior details.

11. Moving the mold blocks:

Select **Insert, Features, Move/Copy Bodies**.

For Bodies to Move/Copy, select the **upper block**.

Enter **6.00in** in the **Delta Y** section.

Click **OK**.

Click Move/Copy Bodies again.

Move the lower block along the **Delta Y** direction **-5.50in.**

Click **OK**.

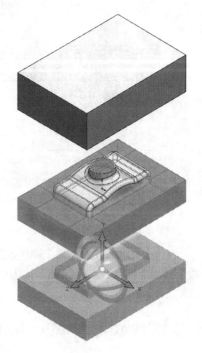

Optional:
Expand the Solid Bodies folder, right click the **Body-Move/Copy1** (the Cavity block) and select **Change Transparency**.

12. Saving your work:

Select **File, Save As**.

Enter **Lesson 10_Exercise_Completed** for the file name.

Click **Save**.

Close all documents.

Chapter 11: Surface Repairs and Patches

This exercise demonstrates several different methods to repair the faulty surfaces while showing us the functions of the frequently used surface commands.

1. Opening a part document:

Select **File, Open**.

Open a part document named:
Repair Surfaces.sldprt

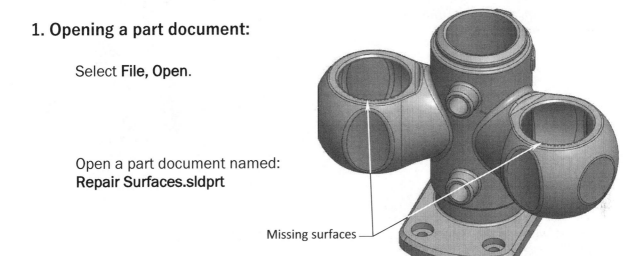

Missing surfaces

Notice the missing surfaces in the first image. The two surfaces used to create the geometry of the holes are missing.

Missing surfaces

Rotate the model and examine the back side. The surfaces at upper left corner of the keyway are missing.

A surface on the raised feature is also missing.

2. Creating the 1st Lofted Surface:

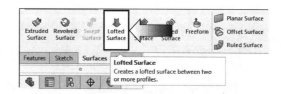

First, we will patch the openings of the holes.

Change to the **Isometric** view (Control+7).

Press the **Down Arrow** <u>3 times</u>. (The default rotate angle is 15°.) Your model should look similar to the one shown on the right.

Switch to the **Surfaces** tab and select the **Lofted Surface** command.

For Loft Profiles, select the <u>upper</u> and <u>lower edges</u> of the hole as indicated. (Do not select the edges of the chamfer.)

The 2 connectors appear in the preview graphics indicating a lofted surface is being created.

Select 2 edges

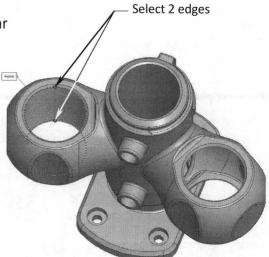

Drag the connectors if needed to keep them vertical to each other. This will prevent the surface from twisting.

Click **OK**.

3. Creating the 2ⁿᵈ Lofted Surface:

Repeat step number 2 and
create the second lofted surface
to repair the hole on the right side.

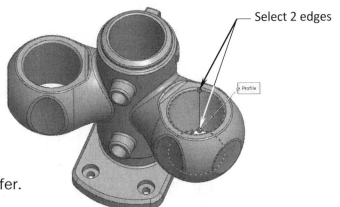

Select 2 edges

Be sure to select the two edges of
the hole, not the edges of the chamfer.

Move the connectors, if needed, to keep them vertical to each other and as close
as possible. This will prevent the lofted surface from twisting or distorting.

4. Patching the Raised Feature:

Press the **Space Bar** to display the Orientation dialog box.

Select the named view: **Raised Feature** to rotate the model to the orientation
shown below. (There are 3 Named-Views that have been saved earlier.)

Raised Feature

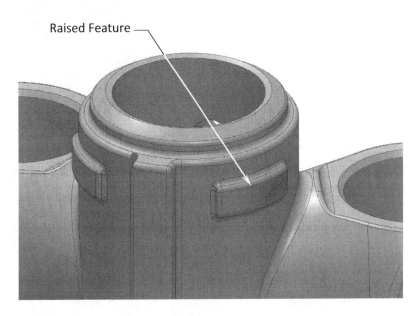

Select the **Filled Surface** command from the **Surfaces** tab (arrow).

The Filled Surface command is used to create a patch around a non-planar boundary.

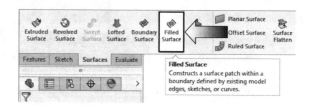

For Patch Boundary, select the **4 edges** in the middle of the raised feature as noted.

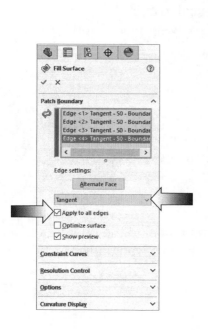

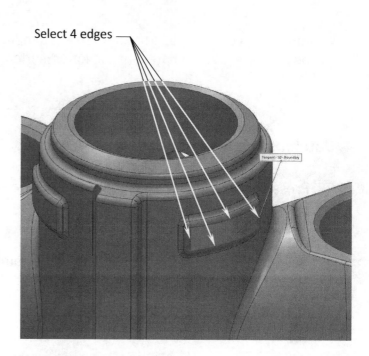

Select 4 edges

Change the Curvature Control to **Tangent** (arrow).

Enable the **Apply to All Edges** checkbox (arrow) and disable the **Optimize Surface** option.

Click **OK**.

5. Patching the corner of the keyway:

Press the **Space Bar** to display the Orientation dialog box.

Select the named view: **Corner of Keyway** to rotate the model to the orientation shown below.

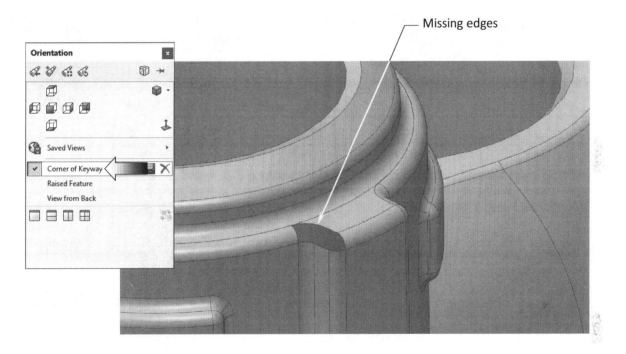

The edges around this corner are missing.

We will take a look at several different methods to fill the opening. Each method will demonstrate the use of different surfacing tool. Some of them may produce an acceptable result while some others may not.

Let us try the Filled Surface command first.

6. Using the Filled Surface command:

Use the Filled Surface command to construct a surface patch with any number of sides within a boundary defined by existing model edges, sketches, or curves. It is also used to fill gaps in a surface model.

Click the **Filled Surface** command and select the **7 edges** as indicated.

For Curvature Control, select the **Tangent** option.

Select 7 edges

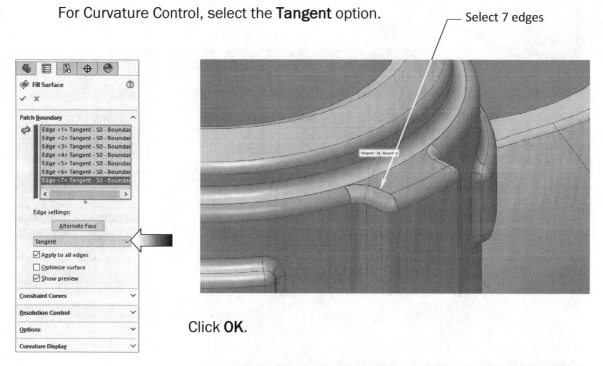

Click **OK**.

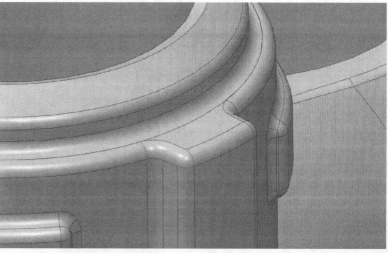

The result appears acceptable, but we will look at the other methods and attempt to produce better results.

The next option is to use the Lofted-Surface command.

7. Using the Lofted Surface command:

Suppress the **Surface-Fill2** feature (arrow).

Select the **Lofted Surface** command from the **Surfaces** tab.

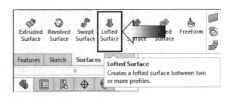

Right-click one of the edges of the openings and select **SelectionManager**. (SelectionManager combines the selected edges into a single group.)

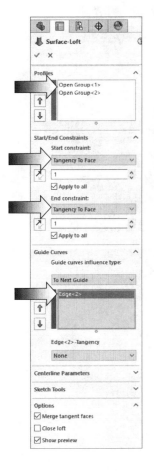

For Group1 of the Loft Profile, select the **top 3 edges** and click the **green checkmark**.

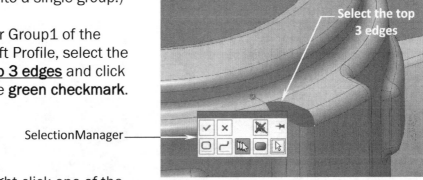

SelectionManager

Right-click one of the other edges and enable the **SelectionManager**.

For Group2 of the Loft Profile, select the **bottom 3 edges** & click the **green checkmark**.

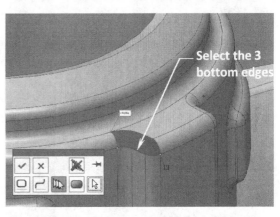

For Guide Curve, select the **curved edge** on the **left side** of the opening.

Expand the Start/End Constraint section and set both options to: **Tangency to Face**.

Click **OK**.

The patch appears better than the surface fill that was created in the last step. Let us try another method.

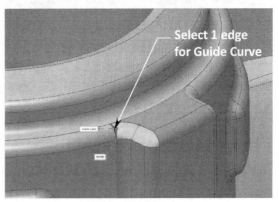

8. Using the Boundary Surface command:

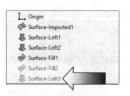

Suppress the Surface-Loft3 feature (arrow).

Select the Boundary Surface command from the Surfaces tab.

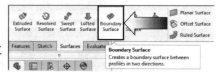

For Direction 1, use SelectionManager and select the **upper 2 edges**; click the green checkmark.

Use the SelectionManager to select the bottom **2 edges** to complete Direction 1.

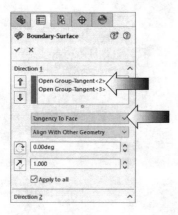

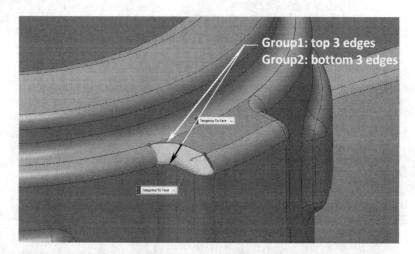

Select **Tangency to Face** for both groups.

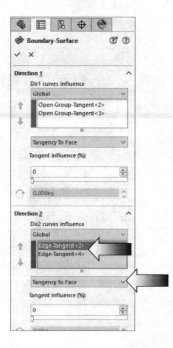

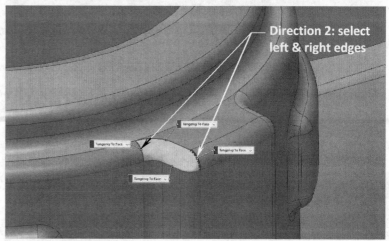

For Direction 2, select the **2 edges** on the left and right side of the patch opening. Select **Tangency to Face** for both edges.

Click **OK**. The boundary patch is acceptable this time.

9. Creating the additional curves:

Suppress the Boundary-Surface1 feature (arrow).

Additional curves are needed in order to achieve a smooth patch while accurately filling the corner of the keyway.

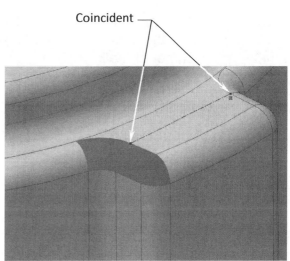

Switch to the Sketch tab and select 3D Sketch.

Sketch a centerline that is coincident to the vertices on both sides of the keyway.

Sketch a Spline that connects the two vertices as shown.

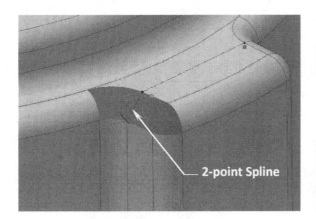

Add a Tangent relation between the spline and the vertical edge as indicated.

Add another Tangent relation between the spline and the centerline.
This is Curve-1 which will be used to define the boundary of the upper left corner.

Sketch another **2-point Spline** that connects the 2 vertices as noted.

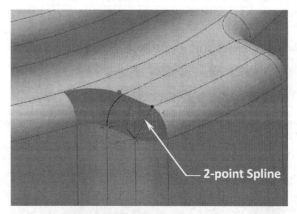

2-point Spline

Add **2 Tangent** relations between the spline and the edges on both sides of the keyway.

This is Curve-2 which will be used to define the boundary of the upper right corner of the keyway.

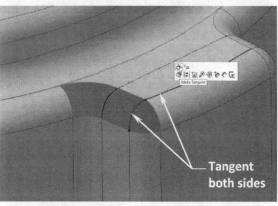

Tangent both sides

Exit the 3D sketch.

10. Creating a Loft with Guide Curves:

Select the **Lofted Surface** command.

For Loft Profile, use **SelectionManager** to select the top 3 edges as Group-1, and select the bottom 3 edges as Group-2.

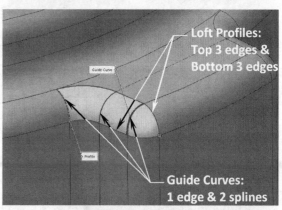

Loft Profiles:
Top 3 edges &
Bottom 3 edges

Guide Curves:
1 edge & 2 splines

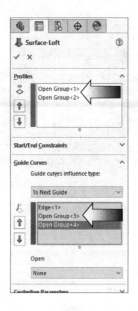

For Guide Curves, select the curved edge on the left side and the 2 splines as indicated.

The Tangent relations have already been added in the 3D Sketch, so no other constraints are needed here.

Click **OK**.

The result shows the best patch in comparison to the others. Now let us move on to other patches.

11. Removing features:

Change to the **Isometric** orientation (Control+7).

The lower cylindrical boss feature needs to be removed. One way to achieve this is to use the Delete Face command.

Zoom in on the cylindrical boss feature.

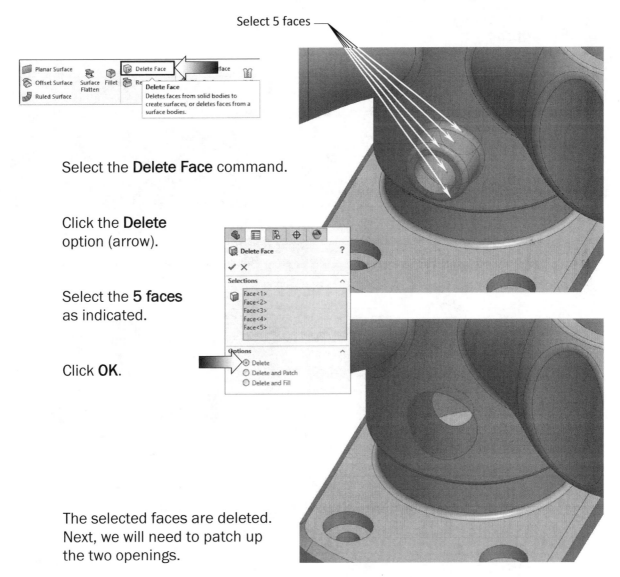

Select the **Delete Face** command.

Click the **Delete** option (arrow).

Select the **5 faces** as indicated.

Click **OK**.

The selected faces are deleted. Next, we will need to patch up the two openings.

12. Deleting Hole(s):

The option Delete Hole(s) removes and heals the edges of an opening from a surface.

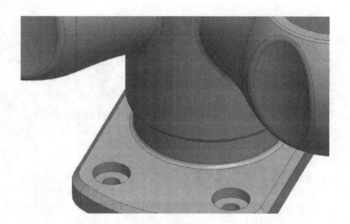

Hold the <u>Control</u> key and select the **2 edges** as noted in the image.

Select 2 edges

Press the **Delete** key on the <u>keyboard</u>.

The Choose Option dialog box appears. Select the **Delete-Hole(s)** option (arrow).

Click **OK**.

The 2 openings are removed and healed at the same time.

13. Removing features:

Press the **Spacebar** and select the named view: **View from Back**.

The two lower rectangular features also need to be removed.

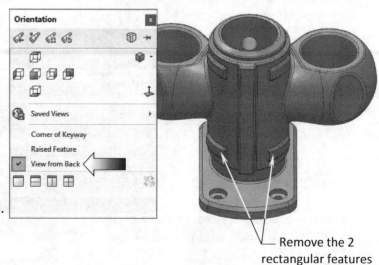

Remove the 2 rectangular features

Let us use the Delete Face command once again.
Zoom in on one of the rectangular features.

Select the **Delete Face** command.

Click the **Delete** option (arrow).

Select **27 faces** of the rectangular feature, including the fillets around it.

Click **OK**.

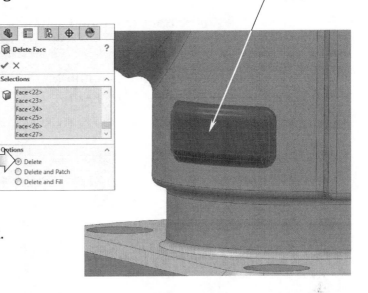

Select 27 faces

Repeat the step above and remove the second rectangular feature on the other side.

The rectangular features are removed, leaving 2 openings that need to be filled.

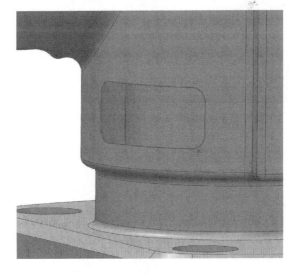

Use the **Delete Hole(s)** option to fill the 2 openings (select the edges of the openings, press Delete on the keyboard).

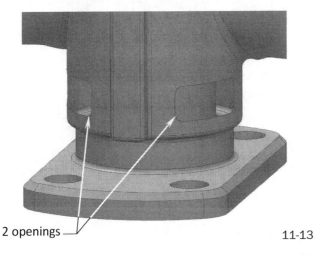

2 openings

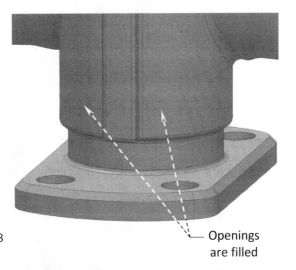

Openings are filled

14. Knitting the surfaces:

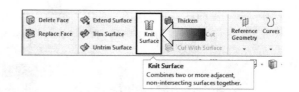

Select the **Knit Surface** command.

For Selections, expand the FeatureManager tree and select the **5 surfaces** inside the **Surface-Bodies folder** (arrow).

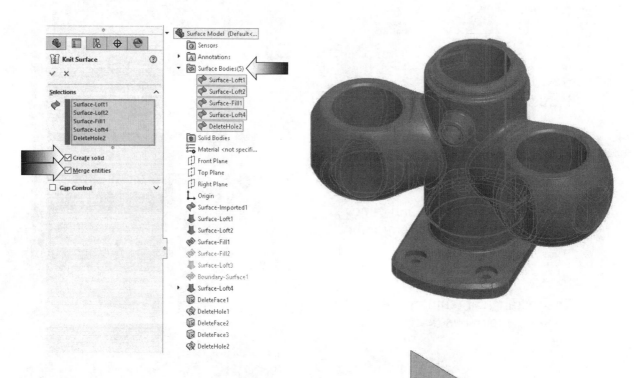

Enable the 2 checkboxes for **Create Solid** and **Merge Entities**.

Click **OK**.

All surfaces are knitted into a single body and at the same time, converted into a solid model.

Create a **Section View** to verify the solid material of the model. Cancel the Section View when done verifying.

15. Assigning material:

Right-click the **Material** option on the Feature Manager and select: **Cast Alloy Steel** (arrow).

This material has the density of:
0.264 pounds per cubic inch.

16. Calculating the mass:

Click the **Evaluate** tab and select the **Mass Properties** command (arrow).

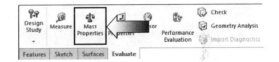

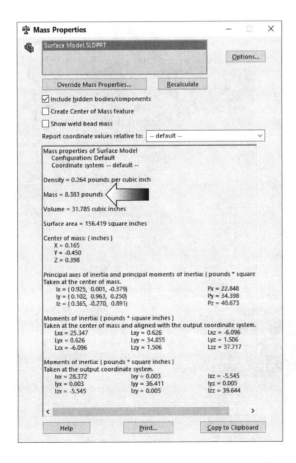

Locate the mass of your model and enter it here:

_____ pounds

17. Saving your work:

Select **File, Save As.**

Enter **Repairing Surfaces (Completed).sldprt**

Click **Save.**

Exercise: Surface Repair & Patches

The exercises give you an opportunity to apply what you have learned from the lessons. The instructions are limited to allow you to try things out using your own techniques, at your own pace.

1. Opening a part document:

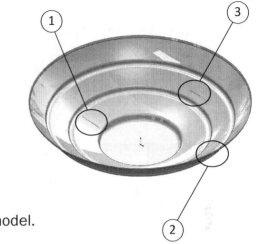

Browse to the Training Folder and open a part document named:
Lesson 11_Exercise.sldprt.

There are some cracks and slivers in the model. They are labeled as 1, 2, and 3. Those areas need to be trimmed and patched prior to converting into a solid model.

2. Filling the opening 1:

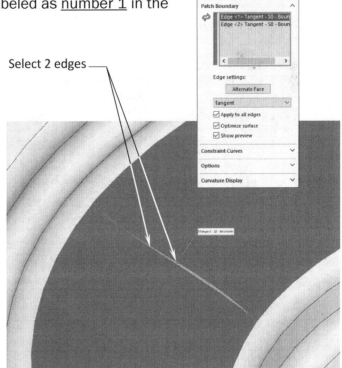

Zoom in closer to the **gap** labeled as <u>number 1</u> in the image above.

Click **Filled Surface.**

Select the <u>2 edges</u> as indicated.

Change the Curvature Control to **Tangent.**

Enable the checkboxes:
Apply to All Edges
Optimize Surface

Click **OK.**
The gap is filled.

Select 2 edges

3. Patching the upper portion of the opening 2:

Let us try another method for patching up the openings.

Click **Lofted Surface**.

For Loft Profiles, select the **2 vertical edges** as shown.

For Guide Curves, select the **2 splines** in the 3D Sketch.

Click **OK**.

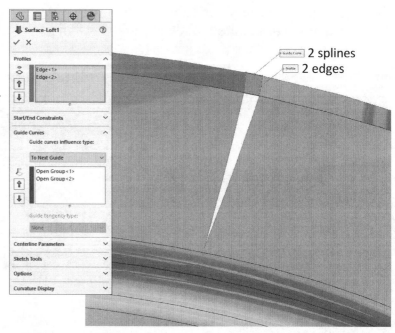

2 splines
2 edges

4. Patching the lower portion of the opening 2:

Click **Filled Surface**.

Select the **3 edges** as noted.

Change the Curvature Control to **Contact**.

Enable the check-boxes:

    Apply to All Edges
    Optimize Surface

Click **OK**.

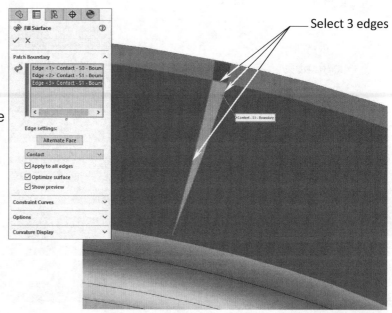

Select 3 edges

5. Deleting the sliver:

There is an extra surface floating in the back of the surface model. We need to delete it.

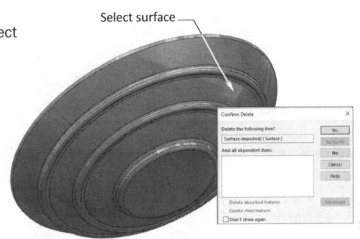

Select surface

Rotate the model and select the **surface** as noted.

Press the <u>Delete key</u> on the keyboard.

Click **Yes** to confirm the delete.

6. Patching the opening 3:

Zoom in on the **gap** labeled as <u>number 3</u>.

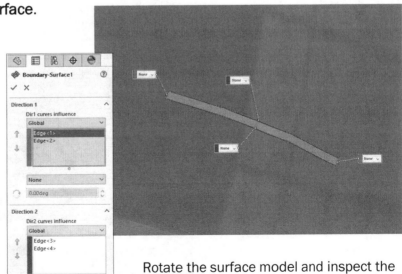

Click **Boundary Surface**.

For Direction 1, select the **2 short vertical edges** (green tags).

For Direction 2, select the **2 long curved edges** (purple tags).

Click **OK**.

Rotate the surface model and inspect the back side to ensure there are no gaps left in the model.

7. Knitting the surfaces:

In order to thicken the model, all surfaces must be knitted into a single surface.

Click **Knit Surface**.

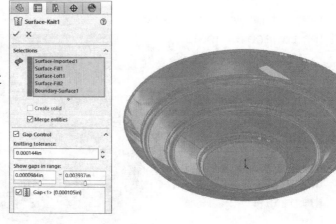

Expand the **Surface Bodies folder** and select all **5 surfaces** inside.

Enable the **Gap Control** checkbox and click the **Gap** checkbox to allow the gap to be healed.

Click **OK**.

8. Thickening the surface body:

Click **Thicken**.

For Thicken Parameters, select the **surface** model from the graphics area.

For Thickness Direction click **Both Sides**.

For Wall Thickness, enter **.040in**.

Click **OK**.

9. Saving your work:

Select **File, Save As**.

Enter **Lesson 11_Exercise_Completed** for the file name.

Click **Save**.

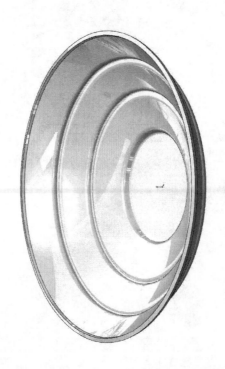

Chapter 12: Rendering with PhotoView 360

PhotoView 360 is a SOLIDWORKS add-in that produces photo-realistic renderings of SOLIDWORKS models. The rendered image incorporates the appearances, lighting, scene, and decals included with the model.

This lesson will walk us through a few simple steps to enlighten your design and make it look more attractive when sharing it with others.

1. Opening an assembly document:

Select **File, open**.

Browse to the Training Folder and open the assembly document named: **Space Station_ Rendering.sldasm**

This is a fairly large assembly document. It contains one Space Station component, 24 Jet Planes and 24 Helicopters.

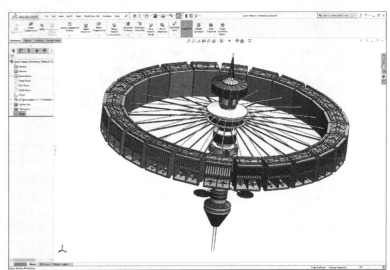

2. Enabling PhotoView 360:

Select **Tools, Add-Ins**.

Enable the checkbox for **PhotoView 360**.

<u>Note</u>: PhotoView is available with SOLIDWORKS Pro. or SOLIDWORKS Premium.

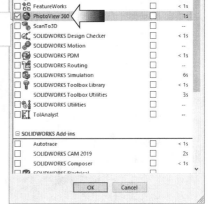

3. Changing the Scene:

The first step is to assign a scene that looks the best for your design. For this example, we will use the scene called **3 Point Beige**, available in the <u>Scene</u> folder.

From the **Task Pane** click the **Appearances** tab and expand the **Scene** and **Basic Scene** folder.

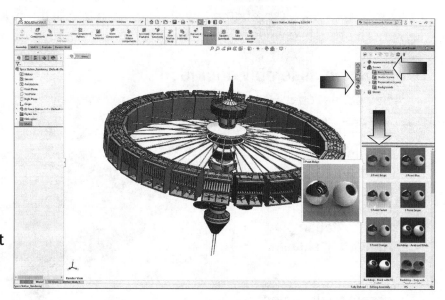

Double click on the **3 Point Base** scene to apply it to the background.

The background changes to a beige color. The default lights for this scene are loaded in the View Scene, Lights, and Cameras tab.

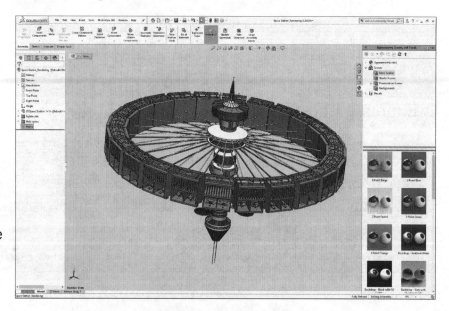

Expand the Lights folder to change the colors or the positions if necessary.

4. Enabling the Preview Window:

The preview window helps you assess changes quickly before performing a full render.

Click the **PhotoView 360** tab and enable the **Preview Window**.

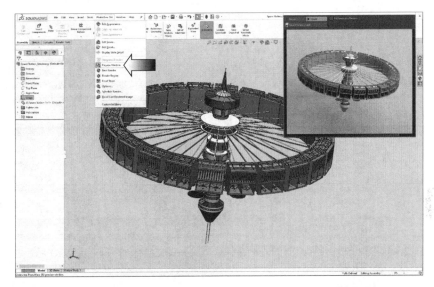

The preview updates continuously as you change the model, refining the model incrementally. Changes to appearances, decals, scenes, and rendering options update in real time. If you change portions of the model, the preview updates for those portions only, rather than for the entire display.

5. Setting the output image size:

This step sets the width and height of the output image, in pixels. The higher image size will produce higher resolution image, but it increases the file size at the same time.

Click the **PhotoView 360** tab and select **Options**.

Under Output Image Size, select: **1920x1080 (16:9).**

Set the other options as shown.

6. Creating the Rendering:

Click the **PhotoView 360** tab again and select: **Final Render**.

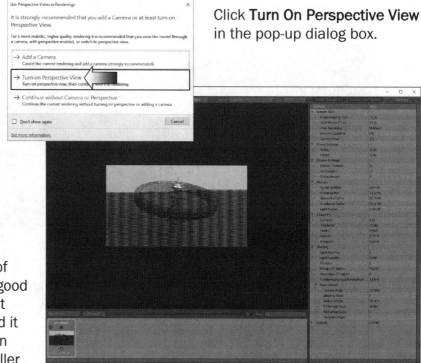

Click **Turn On Perspective View** in the pop-up dialog box.

The resolution of **1920x1080** is good enough for most applications and it looks good when printing on smaller paper size such as **8.5"x11.0"**

When zoomed closer, the image is pixelated as shown in the image on the right.

We will try to improve the image quality by increasing the image size by 2 times.

7. Increasing the image quality:

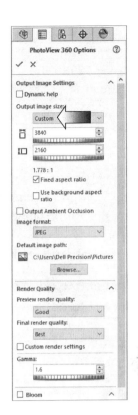

Click **PhotoView 360** tab and select **Options**.

For output Image Size, select **Custom**.

Change the Image Width to **3840**

Change the Image Height to **2160**

Click **OK**.

Click **PhotoView 360, Final Render**.

Since the image size has just been doubled, the render time will take a little bit longer than the last time.

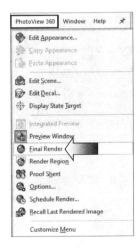

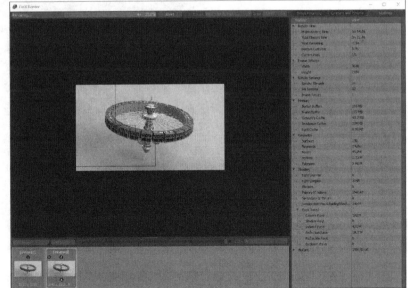

The rendered images can be saved as various windows supported formats such as: .bmp, gif, tif, png, jpg, etc.

The 2nd rendering looks much better than the 1st one.

It is good for printing up to **11.0"x17.0"** paper size.

It is still pixelated when zoomed in much closer.

So, for larger print sizes such as posters or banners, we need to double or even quadruple the image size.

8. Quadrupling the image quality:

Click **PhotoView 360** tab and select **Options**.

For output Image Size, select **Custom**.

Change the Image Width to **7680** and change the Image Height to **4320**

Click **OK**.

Render the image once again. The image quality is now suitable for larger size prints.

9. Saving your work:

Click **Files, Save As**.

Enter **Space Station_Rendering_Finished** for the file name.

Click **Save**.

The Space Station and the Rendering were designed using SOLIDWORKS 2023

Exercise 1: Rendering with PhotoView 360

The exercises give you an opportunity to apply what you have learned from the lessons. The instructions are limited to allow you to try things out using your own techniques, at your own pace.

1. Opening a part document:

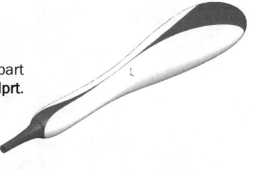

Browse to the Training Folder and open a part document named: **Lesson 12_Exercise.sldprt.**

This model was previously created in lesson 5. We will use a copy of it and create a rendering using PhotoView 360.

2. Enabling PhotoView 360 application:

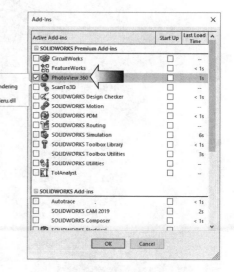

Select **Tools, Add-Ins.**

Enable the **PhotoView 360** checkbox.

A PhotoView 360 drop-down menu is added to the menu bar.

There are 4 main steps to create a rendering of a model:

> **Select the Scene**
>
> **Apply the Appearances**
>
> **Set the Image Output size**
>
> **Run the Final Rendering**

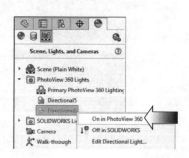

Each scene comes with some pre-set lights; be sure to right-click and turn those lights on.

3. Applying the scene:

Click **Appearances** (arrow) and expand the **Studio Scene**.

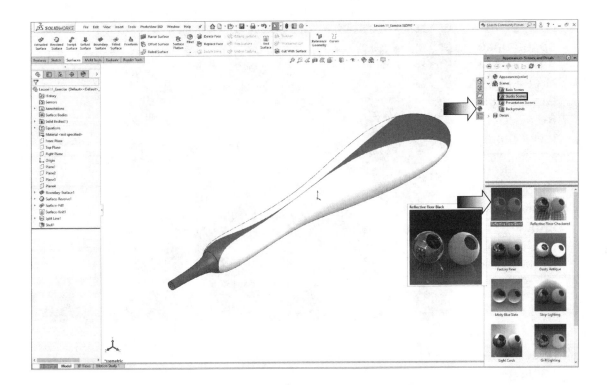

Double-click on the scene named: **Reflective Floor Black**.

A black color and a mirror floor is added to the background of the model. This scene provides a realistic light source, including illumination and reflections, requiring less manipulation of lighting.

The objects and lights in a scene can form reflections on the model and can cast shadows on the floor.

(Double-click on Primary-PhotoView 360 Lights to modify the light settings.)

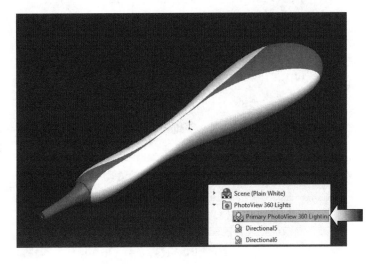

4. Applying Appearances:

Expand the **Plastics** folder and click the **High Gloss** folder to see its content.

Locate the
appearance
named:
Green High
Gloss Plastic
from the
lower pane.

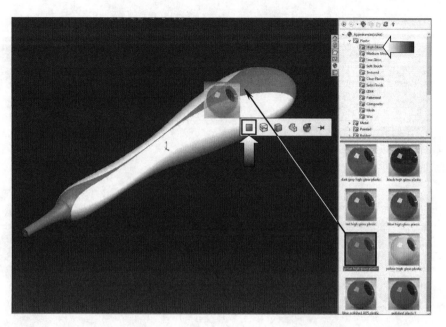

Drag this
appearance
and drop
on the **face**
as indicated
by the arrow.

Click the **Face**
option in the
context-menu.

5. Setting the output image size:

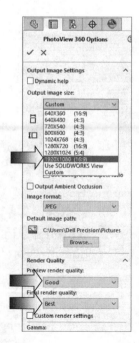

For this exercise, we will use one of the default image sizes
provided in the drop-down menu.

Click the PhotoView 360 drop-down menu and select:
Options.

Select **1920 X 1080** under the Output Image Size list.

Set the Render Quality to:

Good for Preview Render.

Best for Final Render Quality.

6. Creating the final rendering:

Select **Final Render** from the PhotoView 360 drop-down menu.

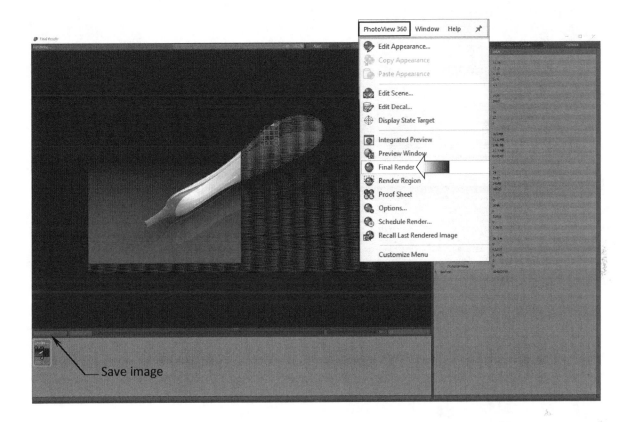

The rendering can be saved as JPG, TIG, PNG, etc. after the rendering is completed.

7. Saving your work:

Select **File, Save As**.

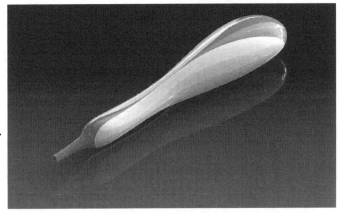

Enter **Lesson 12_Exercise_ Completed** for the file name.

Click **Save**.

Exercise 2: Bug-Bot Rendering

1. Opening a part document:

Browse to the Training Folder and open a part
document named: **Bug_Bot Appearances.sldprt**

Click **NO** in the FeatureWorks dialog box.

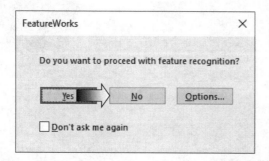

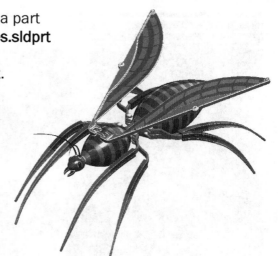

2. Changing the face colors:

Use the Selection Filters to precisely select specific types of items in the
graphics area such as faces, edges, bodies, etc.

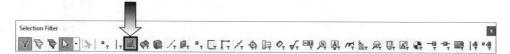

Press the **F5** function key to enable the **Section Filter** toolbar. It usually appears
at the lower left side of the screen.

Enable the **Filter Faces** button (arrow).

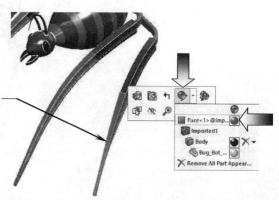

Click this face and select: _____
Appearances, Face

Click the face as indicated and select:
Appearances, Face (arrow).

Select the **White** color from the Color Palette, in the FeatureManager tree.

Select all brown color faces of the legs as indicated. (The option Select-Tangency works well for this portion of the legs.)

Your computer may slow down a little if there are too many faces selected at the same time.

Select all brown color faces

Click **OK**.

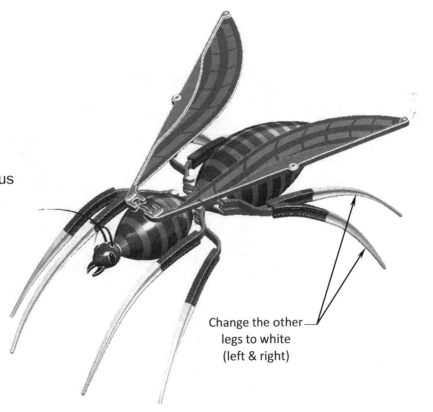

Repeat the previous step and change the color of the other legs, also to **white** color.

Change the other legs to white (left & right)

Click **OK**.

3. Changing the color of the body and the wings:

Click one of the brown color stripes and select **Appearances, Face**.

Select the **White** color from the palette.

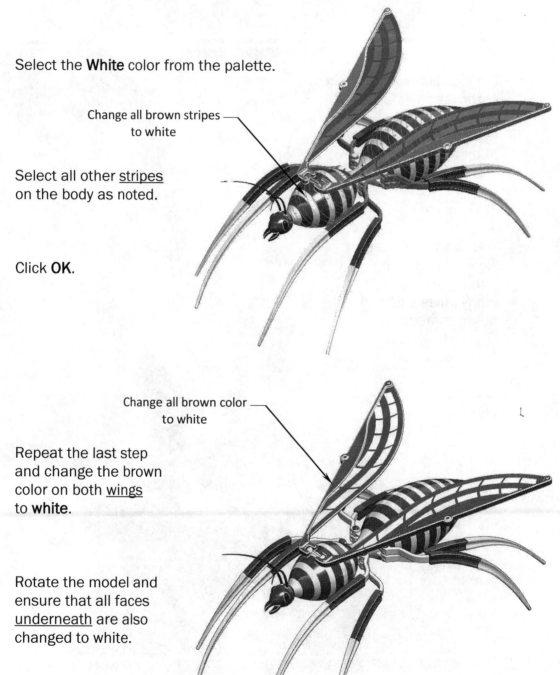

Change all brown stripes to white

Select all other <u>stripes</u> on the body as noted.

Click **OK**.

Change all brown color to white

Repeat the last step and change the brown color on both <u>wings</u> to **white**.

Rotate the model and ensure that all faces <u>underneath</u> are also changed to white.

Click **OK**.

4. Creating the final rendering:

Select **PhotoView 360, Options**.

For Output Image Size, select **1920X1080** from the drop-down list (arrow).

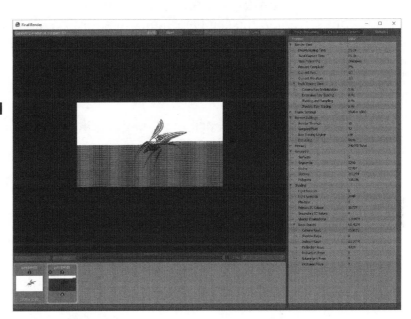

Create the Final Rendering using the the parameters shown in the dialog box.

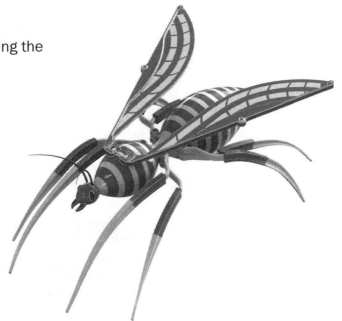

5. Saving your work:

Select **File, Save As**.

Enter: **Bug_Bot Appearances Completed.sldprt** for the file name.

Click **Save**.

Using Display States

Configurations are used to capture and save multiple versions of the part in the same document. Display States are used to create and save the Visual-Properties within a model or assembly such as visibility, appearance, display mode, or transparency of components. Both Configurations and Display States are stored on the ConfigurationManager tree. By default, every configuration comes with a display state and more can be added at any time.

1. Opening an assembly document:

Open an assembly document named: **Display States Assembly.sldasm**

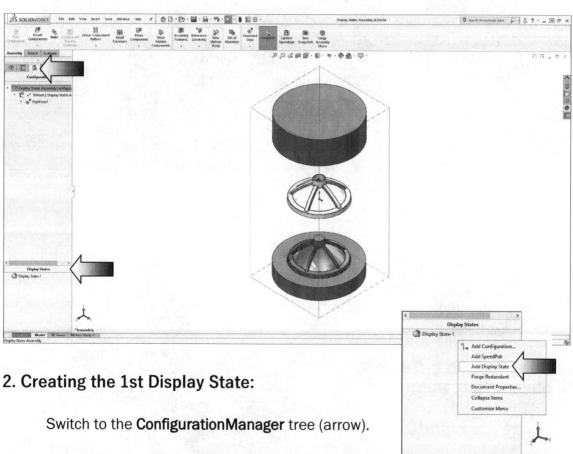

2. Creating the 1st Display State:

Switch to the **ConfigurationManager** tree (arrow).

Right-click the default Display State-1 and select: **Add Display State** (arrow).

Change the name of the new Display State to: **Blue Part – Core & Cavity Transparent** (arrow).

3. Changing the appearances:

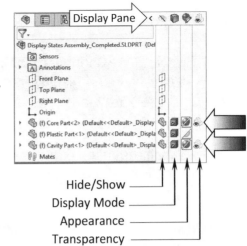

Switch back to the **FeatureManager** tree and expand the **Display Pane** (arrow).

Locate the component named: **Core Part** and click the **Transparent** button on the 4th column (Transparency Column).

Also make the **Cavity Part Transparent** the same way (arrow).

Hide/Show
Display Mode
Appearance
Transparency

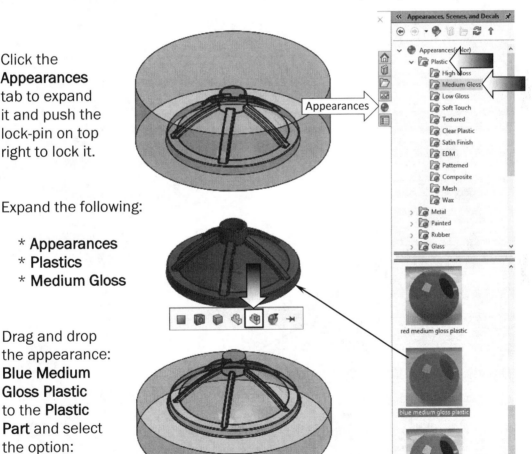

Click the **Appearances** tab to expand it and push the lock-pin on top right to lock it.

Expand the following:

* **Appearances**
* **Plastics**
* **Medium Gloss**

Drag and drop the appearance: **Blue Medium Gloss Plastic** to the **Plastic Part** and select the option: **Component** (arrow).

Appearances

red medium gloss plastic

blue medium gloss plastic

green medium gloss plastic

A new Display State is created.
In this display state, the appearance of the Core and Cavity parts are changed to transparent and the Plastic Part is changed to the Blue color.

4. Creating the 2nd Display State:

Switch to the **ConfigurationManager** tree.
Right-click in the Display States area and select:
Add Display State (arrow).

Change the name of the new Display State to:
Green Part - Core & Cavity Wireframe (arrow).

Switch back to the **FeatureManager** tree.
Locate the component **Core Part** and click
the **Display Mode** button on the 2nd column.
Select the **Wireframe** option (arrow).

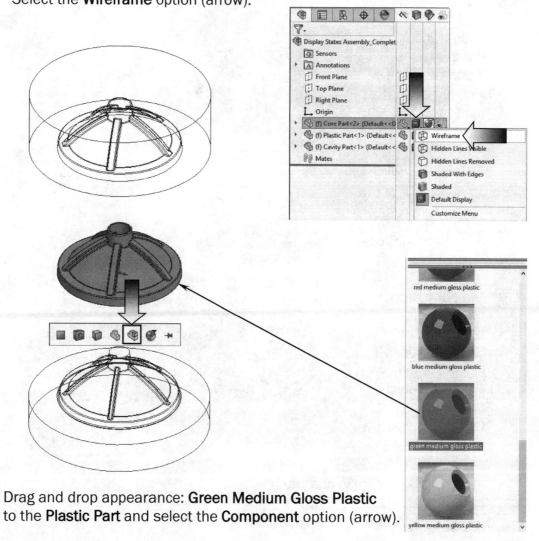

Drag and drop appearance: **Green Medium Gloss Plastic**
to the **Plastic Part** and select the **Component** option (arrow).

5. Adding the 3rd Display State:

Switch to the **Configuration-Manager** tree.

Right-click in the Display State area and select: **Add Display State**.

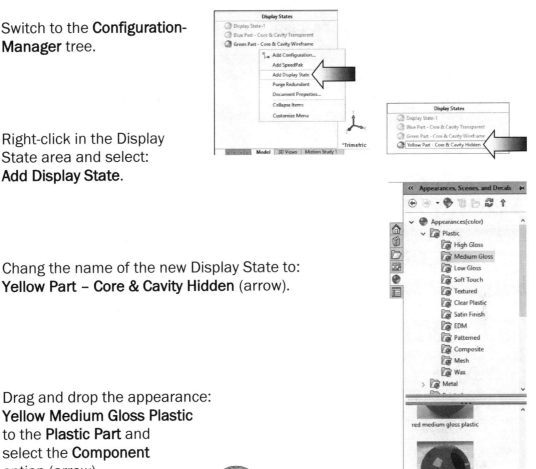

Chang the name of the new Display State to: **Yellow Part – Core & Cavity Hidden** (arrow).

Drag and drop the appearance: **Yellow Medium Gloss Plastic** to the **Plastic Part** and select the **Component** option (arrow).

Switch back to the **FeatureManager** tree.
Locate the 2 components **Core Part** and **Cavity Part** and click the **Hide/Show** button on the first column.

Including the default, there should be a total of 4 Display States at this point.

6. Toggling between Display States:

The Display States can be toggled back and forth by double-clicking on their names.

Double-click each Display State to verify the appearance changes between the components.

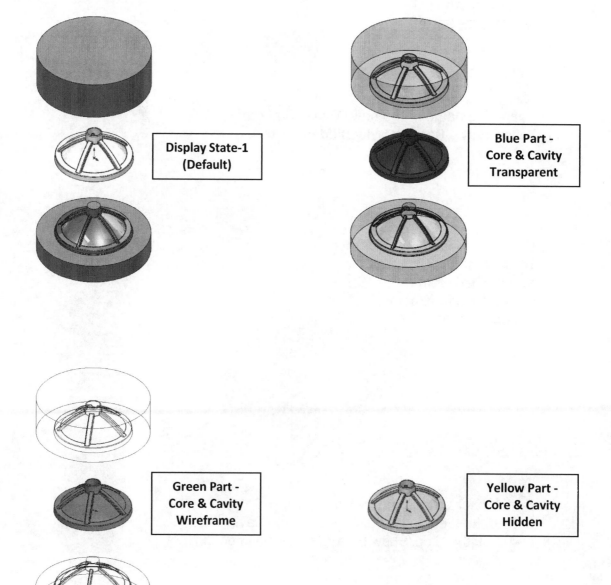

Display State-1 (Default)

Blue Part - Core & Cavity Transparent

Green Part - Core & Cavity Wireframe

Yellow Part - Core & Cavity Hidden

Optional:

1. *Drag the FeatureManager tree **Split Handle** down approximately halfway.*

2. *Create a new configuration named: **Partial Section**.*

3. *Sketch a **Corner Rectangle** (7.00in X 7.00in) on the Top face of the Cavity Part.*

4. *Make an **Assembly Feature** Extruded-Cut Through-All components.*

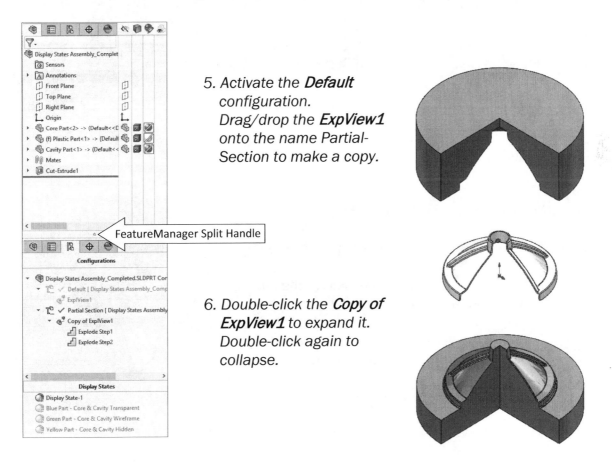

5. *Activate the **Default** configuration. Drag/drop the **ExpView1** onto the name Partial-Section to make a copy.*

6. *Double-click the **Copy of ExpView1** to expand it. Double-click again to collapse.*

7. Saving your work:

Select **File, Save As**.

Enter: **Display States Assembly_Completed.sldasm**
for the file name.

Click **Save**.

Final Exam
Hybrid Modeling

Instructions will be limited so that you can apply what you have learned from the previous lessons to create the required features for this model.

1. Opening a part document:

Browse to the Training Folder and open a part document named: **Advanced Surfaces_Final Exam.sldprt**

There are no features on the FeatureManager tree except for a single surface body of the Detergent Container.

You will need to perform the following tasks:

1. Convert the surface model into a solid body.

2. Create 2 rectangular recesses for the labels, both front and back.

3. Create the grip-features on the left and right sided of the detergent container.

4. Add fillets to all recesses and grip-features.

5. Add a constant wall thickness of .020" to the model.

6. Assign material and calculate the overall mass of the solid model.

Notes:

Follow the instructions in this lesson only when needed. You can use any methods that you want to create the required features, as long as they have the right dimensions and constraints; and most importantly, they must have the correct overall mass as the answer to the final question.

2. Adding the planar surfaces:

In order to convert to a solid body all surfaces must form a closed volume. The 2 openings on top of the container must be closed.

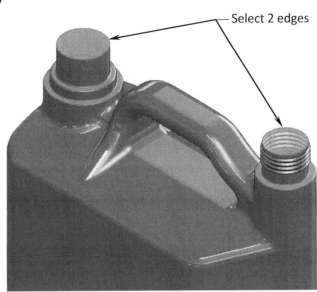

Select 2 edges

Click **Planar Surface.**

Select the **2 circular edges** as noted.

Click **OK.**

3. Converting to solid:

Click **Knit Surface.**

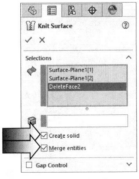

For Selections, select all **3 surfaces** either from the Surface Bodies folder or directly from the graphics area.

Enable the checkboxes shown in the dialog.

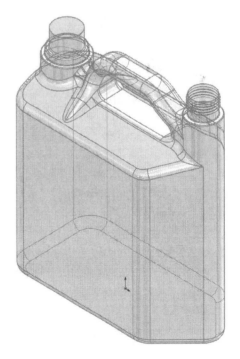

Click **OK.**
(Create a section view to verify the solid body).

4. Creating a recess for the label:

Select the <u>front face</u> as indicated and open a **new sketch**.

Sketch face ───

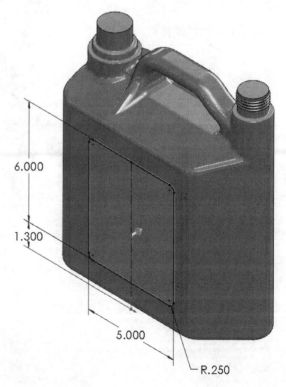

Sketch a **Rectangle** and add the size and location dimensions.

Add a **Sketch Fillet** of .250in. to 4 corners.

Switch to the **Features** tab.

Click **Extruded Cut**.

For Direction 1, use the default **Blind** type.

For Depth, enter .050in.

Click **OK**.

5. Creating the grip-sketch:

Select the <u>front face</u> of the container and open a **new sketch**.

Sketch face ────

Sketch **2 Rectangles** on the left and right sides as shown.

Add the **Sketch Fillets** of **.250in.** to 4 corners.

Switch to the **Features** tab.

Click **Curves, Split Line**.

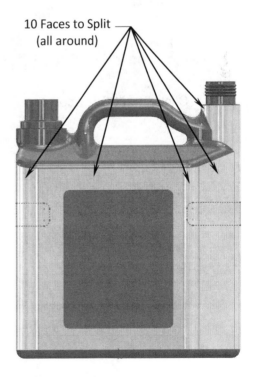

Front Plane

1.000

R.250

5.750

1.500

2.375

For Type of Split, select **Projection**.

For Split Sketch, select **Sketch2** (it should be selected by default).

For Selections, select **10 faces** all around the model.

Click **OK**.

10 Faces to Split (all around)

Rotate the model and inspect the split. It should go through both sides of the model.

6. Making the 1st surface offset:

Two or more surfaces are needed to create a Lofted-Cut feature.

Switch to the **Surfaces** tab.

Click **Offset Surface**.

For Offset Parameters, select **6 surfaces** on the right.

For Offset Distance, enter **.125in.** and click **Reverse** to place the offset surface on the <u>inside</u>.

Click **OK**.

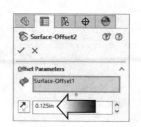

<u>Hide</u> the solid body to see the Offset Surface

7. Making the 2nd surface offset:

Click **Offset Surface**.

For Offset Parameters, select the same **6 surfaces** on the right once again.

For Offset Distance, enter **.125in.**

Do not click Reverse.
Place the 2nd offset surface on the <u>outside</u>.
Click **OK**.

8. Creating the 1st grip-feature:

Switch to the **Features** tab.

Click **Lofted Cut**.

Select the **2 Offset Surfaces**.

The 2 connectors should be on top of each other to prevent the loft from twisting. If needed, drag the connectors to re-position them.

Connectors ────

Click **OK**.

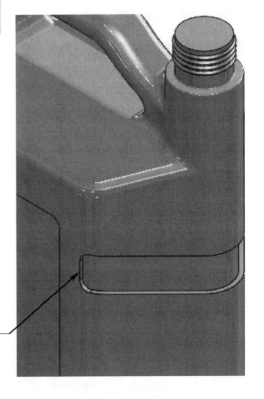

Hide both Surface Offset 1 and 2 to see the lofted cut feature.

Zoom closer and inspect the 1st grip feature. The depth of cut was defined by the distance of each offset surface.

Note: *The Ruled Surface option can also be used to create the cut.*

The depth was determined ──── by the distances between the 2 offset surfaces

9. Making the 3rd offset surface:

Repeat step 6, 7, and 8 to create the grip feature
on the left side.

Click **Offset Surface**.

For Offset Parameters,
select **5 surfaces**
on the left side.

For Offset Distance,
enter **.125in**. and click
Reverse to place the offset
surface on the <u>inside</u>.

Click **OK**.

10. Making the 4th offset surface:

Click **Offset Surface**.

For Offset Parameters,
select the same **5 surfaces**
on the left side once again.

For Offset Distance, enter
.125in.

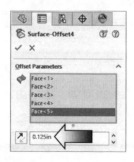

Do not click Reverse.
Place the 2nd offset
surface on the <u>outside</u>.

Click **OK**.
<u>Hide</u> the solid body to see the Offset Surface.

11. Creating the 2nd grip-feature:

Switch to the **Features** tab.

Click **Lofted Cut.**

Select the **2 Offset Surfaces**.

The 2 connectors should
be next to each other
to prevent the loft from
twisting. If needed, drag
the connectors to re-
position them.

Click **OK.**

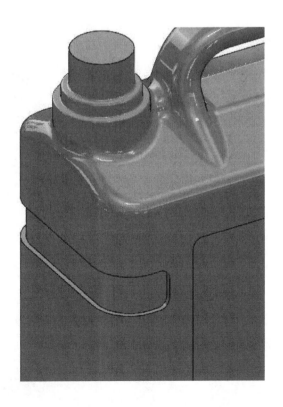

<u>Hide</u> both Surface Offset 3 and 4
to see the lofted cut feature.

Zoom closer and inspect the 2nd grip
feature. The depth of cut was defined
by the distance of each offset surface.

Both Grip-Features will be patterned
6 times in the next couple of steps.

12. Mirroring the label recess:

Click the **Features** tab.

Select **Mirror**.

For Mirror Face/Plane, select **Front** plane.

For Features to Mirror, select the **Cut-Extrude3**, or the rectangular label.

Click **OK**.

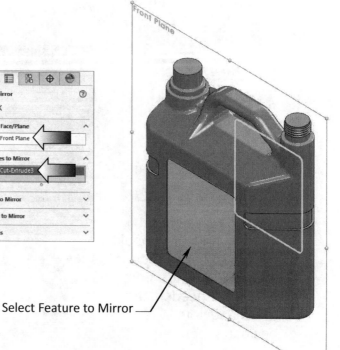

Select Feature to Mirror

13. Patterning the grip-features:

Click **Linear Pattern**.

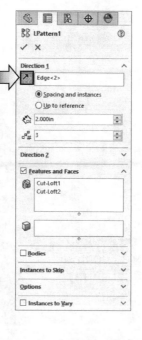

For Direction 1, select the **vertical edge** of the rectangular label as noted.

For Spacing, enter **2.00in**.

For number of Instances, enter **3**.

For Features to pattern, select the **2 Lofted Cuts**.

Click **OK**.

Pattern Direction

14. Adding Fillets:

Click **Fillet**.

Use the default **Constant Size** type.

For Radius, enter
.025in. and select
the **edges** of the
rectangular label,
both front and back.

Click **OK**.

R.025 front & back

R.060 left & right

Also add a fillet of **.060in** to the **edges**
of the **grip-features**, both left and
right sides as indicated.

Use the shortcut
Select-Tangency to ensure
all edges of the grip-features are selected.

Select 2 faces

15. Shelling the model:

Click **Shell**.

For thickness, enter
.025in.

For Faces to Remove,
select the **2 upper faces**
as indicated.

Click **OK**.

Create a section view to verify the wall thickness.

16. Assigning material:

Right-click the material
option and select:
Edit Material.

Expand the **Plastics**
folder and select:
PBT General Purpose
(Polybutylene
Terephthalate).

Click **OK.**

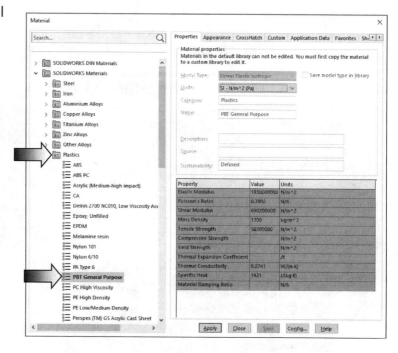

17. Calculating the final mass:

Switch to the **Evaluate** tab.

Click **Mass Properties.**

Enter the mass of your model here:

_____ pounds.

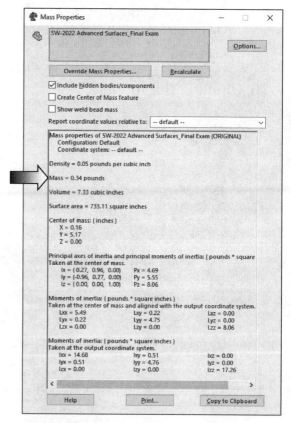

Glossary

Absorbed

A feature, sketch, or annotation that is contained in another item (usually a feature) in the FeatureManager design tree. Examples are the profile sketch and profile path in a base-sweep, or a cosmetic thread annotation in a hole.

Align

Tools that assist in lining up annotations and dimensions (left, right, top, bottom, and so on). For aligning parts in an assembly.

Alternate position view

A drawing view in which one or more views are superimposed in phantom lines on the original view. Alternate position views are often used to show range of motion of an assembly.

Anchor point

The end of a leader that attaches to the note, block, or other annotation. Sheet formats contain anchor points for a bill of materials, a hole table, a revision table, and a weldment cut list.

Annotation

A text note or a symbol that adds specific design intent to a part, assembly, or drawing. Specific types of annotations include note, hole callout, surface finish symbol, datum feature symbol, datum target, geometric tolerance symbol, weld symbol, balloon, and stacked balloon. Annotations that apply only to drawings include center mark, annotation centerline, area hatch, and block.

Appearance callouts

Callouts that display the colors and textures of the face, feature, body, and part under the entity selected and are a shortcut to editing colors and textures.

Area hatch

A crosshatch pattern or fill applied to a selected face or to a closed sketch in a drawing.

Assembly

A document in which parts, features, and other assemblies (sub-assemblies) are mated together. The parts and sub-assemblies exist in documents separate from the assembly. For example, in an assembly, a piston can be mated to other parts, such as a connecting rod or cylinder. This new assembly can then be used as a sub-assembly in an assembly of an engine. The extension for a SOLIDWORKS assembly file name is .SLDASM.

Attachment point
The end of a leader that attaches to the model (to an edge, vertex, or face, for example) or to a drawing sheet.

Axis
A straight line that can be used to create model geometry, features, or patterns. An axis can be made in several different ways, including using the intersection of two planes.

Balloon
Labels parts in an assembly, typically including item numbers and quantity. In drawings, the item numbers are related to rows in a bill of materials.

Base
The first solid feature of a part.

Baseline dimensions
Sets of dimensions measured from the same edge or vertex in a drawing.

Bend
A feature in a sheet metal part. A bend generated from a filleted corner, cylindrical face, or conical face is a round bend; a bend generated from sketched straight lines is a sharp bend.

Bill of materials
A table inserted into a drawing to keep a record of the parts used in an assembly.

Block
A user-defined annotation that you can use in parts, assemblies, and drawings. A block can contain text, sketch entities (except points), and area hatch, and it can be saved in a file for later use as, for example, a custom callout or a company logo.

Bottom-up assembly
An assembly modeling technique where you create parts and then insert them into an assembly.

Broken-out section
A drawing view that exposes inner details of a drawing view by removing material from a closed profile, usually a spline.

Cavity
The mold half that holds the cavity feature of the design part.

Center mark
A cross that marks the center of a circle or arc.

Centerline

A centerline marks, in phantom font, an axis of symmetry in a sketch or drawing.

Chamfer

Bevels a selected edge or vertex. You can apply chamfers to both sketches and features.

Child

A dependent feature related to a previously built feature. For example, a chamfer on the edge of a hole is a child of the parent hole.

Click-release

As you sketch, if you click and then release the pointer, you are in click-release mode. Move the pointer and click again to define the next point in the sketch sequence.

Click-drag

As you sketch, if you click and drag the pointer, you are in click-drag mode. When you release the pointer, the sketch entity is complete.

Closed profile

Also called a closed contour, it is a sketch or sketch entity with no exposed endpoints, for example, a circle or polygon.

Collapse

The opposite of explode. The collapse action returns an exploded assembly's parts to their normal positions.

Collision Detection

An assembly function that detects collisions between components when components move or rotate. A collision occurs when an entity on one component coincides with any entity on another component.

Component

Any part or sub-assembly within an assembly.

Configuration

A variation of a part or assembly within a single document. Variations can include different dimensions, features, and properties. For example, a single part such as a bolt can contain different configurations that vary the diameter and length.

ConfigurationManager

Located on the left side of the SOLIDWORKS window, it is a means to create, select, and view the configurations of parts and assemblies.

Constraint

The relations between sketch entities, or between sketch entities and planes, axes, edges, or vertices.

Construction geometry

The characteristic of a sketch entity that the entity is used in creating other geometry but is not itself used in creating features.

Coordinate system

A system of planes used to assign Cartesian coordinates to features, parts, and assemblies. Part and assembly documents contain default coordinate systems; other coordinate systems can be defined with reference geometry. Coordinate systems can be used with measurement tools and for exporting documents to other file formats.

Cosmetic thread

An annotation that represents threads.

Crosshatch

A pattern (or fill) applied to drawing views such as section views and broken-out sections.

Curvature

Curvature is equal to the inverse of the radius of the curve. The curvature can be displayed in different colors according to the local radius (usually of a surface).

Cut

A feature that removes material from a part by such actions as extrude, revolve, loft, sweep, thicken, cavity, and so on.

Dangling

A dimension, relation, or drawing section view that is unresolved. For example, if a piece of geometry is dimensioned, and that geometry is later deleted, the dimension becomes dangling.

Degrees of freedom

Geometry that is not defined by dimensions or relations is free to move. In 2D sketches, there are three degrees of freedom: movement along the X and Y axes, and rotation about the Z axis (the axis normal to the sketch plane). In 3D sketches and in assemblies, there are six degrees of freedom: movement along the X, Y, and Z axes, and rotation about the X, Y, and Z axes.

Derived part

A derived part is a new base, mirror, or component part created directly from an existing part and linked to the original part such that changes to the original part are reflected in the derived part.

Derived sketch

A copy of a sketch, in either the same part or the same assembly that is connected to the original sketch. Changes in the original sketch are reflected in the derived sketch.

Design Library

Located in the Task Pane, the Design Library provides a central location for reusable elements such as parts, assemblies, and so on.

Design table

An Excel spreadsheet that is used to create multiple configurations in a part or assembly document.

Detached drawing

A drawing format that allows opening and working in a drawing without loading the corresponding models into memory. The models are loaded on an as-needed basis.

Detail view

A portion of a larger view, usually at a larger scale than the original view.

Dimension line

A linear dimension line references the dimension text to extension lines indicating the entity being measured. An angular dimension line references the dimension text directly to the measured object.

DimXpertManager

Located on the left side of the SOLIDWORKS window, it is a means to manage dimensions and tolerances created using DimXpert for parts according to the requirements of the ASME Y.14.41-2003 standard.

DisplayManager

The DisplayManager lists the appearances, decals, lights, scene, and cameras applied to the current model. From the DisplayManager, you can view applied content, and add, edit, or delete items. When PhotoView 360 is added in, the DisplayManager also provides access to PhotoView options.

Document

A file containing a part, assembly, or drawing.

Draft

The degree of taper or angle of a face usually applied to molds or castings.

Drawing

A 2D representation of a 3D part or assembly. The extension for a SOLIDWORKS drawing file name is .SLDDRW.

Drawing sheet

A page in a drawing document.

Driven dimension

Measurements of the model, but they do not drive the model and their values cannot be changed.

Driving dimension

Also referred to as a model dimension, it sets the value for a sketch entity. It can also control distance, thickness, and feature parameters.

Edge

A single outside boundary of a feature.

Edge flange

A sheet metal feature that combines a bend and a tab in a single operation.

Equation

Creates a mathematical relation between sketch dimensions, using dimension names as variables, or between feature parameters, such as the depth of an extruded feature or the instance count in a pattern.

Exploded view

Shows an assembly with its components separated from one another, usually to show how to assemble the mechanism.

Export

Save a SOLIDWORKS document in another format for use in other CAD/CAM, rapid prototyping, web, or graphics software applications.

Extension line

The line extending from the model indicating the point from which a dimension is measured.

Extrude

A feature that linearly projects a sketch to either add material to a part (in a base or boss) or remove material from a part (in a cut or hole).

Face

A selectable area (planar or otherwise) of a model or surface with boundaries that help define the shape of the model or surface. For example, a rectangular solid has six faces.

Fasteners

A SOLIDWORKS Toolbox library that adds fasteners automatically to holes in an assembly.

Feature

An individual shape that, combined with other features, makes up a part or assembly. Some features, such as bosses and cuts, originate as sketches. Other features, such as shells and fillets, modify a feature's geometry. However, not all features have associated geometry. Features are always listed in the FeatureManager design tree.

FeatureManager design tree

Located on the left side of the SOLIDWORKS window, it provides an outline view of the active part, assembly, or drawing.

Fill

A solid area hatch or crosshatch. Fill also applies to patches on surfaces.

Fillet

An internal rounding of a corner or edge in a sketch, or an edge on a surface or solid.

Forming tool

Dies that bend, stretch, or otherwise form sheet metal to create such form features as louvers, lances, flanges, and ribs.

Fully defined

A sketch where all lines and curves in the sketch, and their positions, are described by dimensions or relations, or both, and cannot be moved. Fully defined sketch entities are shown in black.

Geometric tolerance

A set of standard symbols that specify the geometric characteristics and dimensional requirements of a feature.

Graphics area

The area in the SOLIDWORKS window where the part, assembly, or drawing appears.

Guide curve

A 2D or 3D curve used to guide a sweep or loft.

Handle

An arrow, square, or circle that you can drag to adjust the size or position of an entity (a feature, dimension, or sketch entity, for example).

Helix

A curve defined by pitch, revolutions, and height. A helix can be used, for example, as a path for a swept feature cutting threads in a bolt.

Hem

A sheet metal feature that folds back at the edge of a part. A hem can be open, closed, double, or teardrop.

HLR

(Hidden lines removed) a view mode in which all edges of the model that are not visible from the current view angle are removed from the display.

HLV

(Hidden lines visible) A view mode in which all edges of the model that are not visible from the current view angle are shown gray or dashed.

Import

Open files from other CAD software applications into a SOLIDWORKS document.

In-context feature

A feature with an external reference to the geometry of another component; the in-context feature changes automatically if the geometry of the referenced model or feature changes.

Inference

The system automatically creates (infers) relations between dragged entities (sketched entities, annotations, and components) and other entities and geometry. This is useful when positioning entities relative to one another.

Instance

An item in a pattern or a component in an assembly that occurs more than once. Blocks are inserted into drawings as instances of block definitions.

Interference detection

A tool that displays any interference between selected components in an assembly.

Jog

A sheet metal feature that adds material to a part by creating two bends from a sketched line.

Knit

A tool that combines two or more faces or surfaces into one. The edges of the surfaces must be adjacent and not overlapping, but they cannot ever be planar. There is no difference in the appearance of the face or the surface after knitting.

Layout sketch

A sketch that contains important sketch entities, dimensions, and relations. You reference the entities in the layout sketch when creating new sketches, building new geometry, or positioning components in an assembly. This allows for easier updating of your model because changes you make to the layout sketch propagate to the entire model.

Leader

A solid line from an annotation (note, dimension, and so on) to the referenced feature.

Library feature

A frequently used feature, or combination of features, that is created once and then saved for future use.

Lightweight

A part in an assembly or a drawing has only a subset of its model data loaded into memory. The remaining model data is loaded on an as-needed basis. This improves performance of large and complex assemblies.

Line

A straight sketch entity with two endpoints. A line can be created by projecting an external entity such as an edge, plane, axis, or sketch curve into the sketch.

Loft

A base, boss, cut, or surface feature created by transitions between profiles.

Lofted bend

A sheet metal feature that produces a roll form or a transitional shape from two open profile sketches. Lofted bends often create funnels and chutes.

Mass properties

A tool that evaluates the characteristics of a part or an assembly such as volume, surface area, centroid, and so on.

Mate

A geometric relationship, such as coincident, perpendicular, tangent, and so on, between parts in an assembly.

Mate reference

Specifies one or more entities of a component to use for automatic mating. When you drag a component with a mate reference into an assembly, the software tries to find other combinations of the same mate reference name and mate type.

Mates folder

A collection of mates that are solved together. The order in which the mates appear within the Mates folder does not matter.

Mirror

(a) A mirror feature is a copy of a selected feature, mirrored about a plane or planar face. (b) A mirror sketch entity is a copy of a selected sketch entity that is mirrored about a centerline.

Miter flange

A sheet metal feature that joins multiple edge flanges together and miters the corner.

Model

3D solid geometry in a part or assembly document. If a part or assembly document contains multiple configurations, each configuration is a separate model.

Model dimension

A dimension specified in a sketch or a feature in a part or assembly document that defines some entity in a 3D model.

Model item

A characteristic or dimension of feature geometry that can be used in detailing drawings.

Model view

A drawing view of a part or assembly.

Mold

A set of manufacturing tooling used to shape molten plastic or other material into a designed part. You design the mold using a sequence of integrated tools that result in cavity and core blocks that are derived parts of the part to be molded.

Motion Study

Motion Studies are graphical simulations of motion and visual properties with assembly models. Analogous to a configuration, they do not actually change the original assembly model or its properties. They display the model as it changes based on simulation elements you add.

Multibody part

A part with separate solid bodies within the same part document. Unlike the components in an assembly, multibody parts are not dynamic.

Native format

DXF and DWG files remain in their original format (are not converted into SOLIDWORKS format) when viewed in SOLIDWORKS drawing sheets (view only).

Open profile

Also called an open contour, it is a sketch or sketch entity with endpoints exposed. For example, a U-shaped profile is open.

Ordinate dimensions

A chain of dimensions measured from a zero ordinate in a drawing or sketch.

Origin

The model origin appears as three gray arrows and represents the (0,0,0) coordinate of the model. When a sketch is active, a sketch origin appears in red and represents the (0,0,0) coordinate of the sketch. Dimensions and relations can be added to the model origin, but not to a sketch origin.

Out-of-context feature

A feature with an external reference to the geometry of another component that is not open.

Over defined

A sketch is over defined when dimensions or relations are either in conflict or redundant.

Parameter

A value used to define a sketch or feature (often a dimension).

Parent

An existing feature upon which other features depend. For example, in a block with a hole, the block is the parent to the child hole feature.

Part

A single 3D object made up of features. A part can become a component in an assembly, and it can be represented in 2D in a drawing. Examples of parts are bolt, pin, plate, and so on. The extension for a SOLIDWORKS part file name is .SLDPRT.

Path

A sketch, edge, or curve used in creating a sweep or loft.

Pattern

A pattern repeats selected sketch entities, features, or components in an array, which can be linear, circular, or sketch driven. If the seed entity is changed, the other instances in the pattern update.

Physical Dynamics

An assembly tool that displays the motion of assembly components in a realistic way. When you drag a component, the component applies a force to other components it touches. Components move only within their degrees of freedom.

Pierce relation

Makes a sketch point coincident to the location at which an axis, edge, line, or spline pierces the sketch plane.

Planar

Entities that can lie on one plane. For example, a circle is planar, but a helix is not.

Plane

Flat construction geometry. Planes can be used for a 2D sketch, section view of a model, a neutral plane in a draft feature, and others.

Point

A singular location in a sketch, or a projection into a sketch at a single location of an external entity (origin, vertex, axis, or point in an external sketch).

Predefined view

A drawing view in which the view position, orientation, and so on can be specified before a model is inserted. You can save drawing documents with predefined views as templates.

Profile

A sketch entity used to create a feature (such as a loft) or a drawing view (such as a detail view). A profile can be open (such as a U shape or open spline) or closed (such as a circle or closed spline).

Projected dimension

If you dimension entities in an isometric view, projected dimensions are the flat dimensions in 2D.

Projected view
A drawing view projected orthogonally from an existing view.

PropertyManager
Located on the left side of the SOLIDWORKS window, it is used for dynamic editing of sketch entities and most features.

RealView graphics
A hardware (graphics card) support of advanced shading in real time; the rendering applies to the model and is retained as you move or rotate a part.

Rebuild
Tool that updates (or regenerates) the document with any changes made since the last time the model was rebuilt. Rebuild is typically used after changing a model dimension.

Reference dimension
A dimension in a drawing that shows the measurement of an item but cannot drive the model and its value cannot be modified. When model dimensions change, reference dimensions update.

Reference geometry
Includes planes, axes, coordinate systems, and 3D curves. Reference geometry is used to assist in creating features such lofts, sweeps, drafts, chamfers, and patterns.

Relation
A geometric constraint between sketch entities or between a sketch entity and a plane, axis, edge, or vertex. Relations can be added automatically or manually.

Relative view
A relative (or relative to model) drawing view is created relative to planar surfaces in a part or assembly.

Reload
Refreshes shared documents. For example, if you open a part file for read-only access while another user makes changes to the same part, you can reload the new version, including the changes.

Reorder
Reordering (changing the order of) items is possible in the FeatureManager design tree. In parts, you can change the order in which features are solved. In assemblies, you can control the order in which components appear in a bill of materials.

Replace

Substitutes one or more open instances of a component in an assembly with a different component.

Resolved

A state of an assembly component (in an assembly or drawing document) in which it is fully loaded in memory. All the component's model data is available, so its entities can be selected, referenced, edited, and used in mates, and so on.

Revolve

A feature that creates a base or boss, a revolved cut, or revolved surface by revolving one or more sketched profiles around a centerline.

Rip

A sheet metal feature that removes material at an edge to allow a bend.

Rollback

Suppresses all items below the rollback bar.

Section

Another term for profile in sweeps.

Section line

A line or centerline sketched in a drawing view to create a section view.

Section scope

Specifies the components to be left uncut when you create an assembly drawing section view.

Section view

A section view (or section cut) is (1) a part or assembly view cut by a plane, or (2) a drawing view created by cutting another drawing view with a section line.

Seed

A sketch or an entity (a feature, face, or body) that is the basis for a pattern. If you edit the seed, the other entities in the pattern are updated.

Shaded

Displays a model as a colored solid.

Shared values

Also called linked values; these are named variables that you assign to set the value of two or more dimensions to be equal.

Sheet format

Includes page size and orientation, standard text, borders, title blocks, and so on. Sheet formats can be customized and saved for future use. Each sheet of a drawing document can have a different format.

Shell

A feature that hollows out a part, leaving open the selected faces and thin walls on the remaining faces. A hollow part is created when no faces are selected to be open.

Sketch

A collection of lines and other 2D objects on a plane or face that forms the basis for a feature such as a base or a boss. A 3D sketch is non-planar and can be used to guide a sweep or loft, for example.

Smart Fasteners

Automatically adds fasteners (bolts and screws) to an assembly using the SOLIDWORKS Toolbox library of fasteners.

SmartMates

An assembly mating relation that is created automatically.

Solid sweep

A cut sweep created by moving a tool body along a path to cut out 3D material from a model.

Spiral

A flat or 2D helix, defined by a circle, pitch, and number of revolutions.

Spline

A sketched 2D or 3D curve defined by a set of control points.

Split line

Projects a sketched curve onto a selected model face, dividing the face into multiple faces so that each can be selected individually. A split line can be used to create draft features, to create face blend fillets, and to radiate surfaces to cut molds.

Stacked balloon

A set of balloons with only one leader. The balloons can be stacked vertically (up or down) or horizontally (left or right).

Standard 3 views

The three orthographic views (front, right, and top) that are often the basis of a drawing.

StereoLithography

The process of creating rapid prototype parts using a faceted mesh representation in STL files.

Sub-assembly

An assembly document that is part of a larger assembly. For example, the steering mechanism of a car is a sub-assembly of the car.

Suppress

Removes an entity from the display and from any calculations in which it is involved. You can suppress features, assembly components, and so on. Suppressing an entity does not delete the entity; you can unsuppress the entity to restore it.

Surface

A zero-thickness planar or 3D entity with edge boundaries. Surfaces are often used to create solid features. Reference surfaces can be used to modify solid features.

Sweep

Creates a base, boss, cut, or surface feature by moving a profile (section) along a path. For cut sweeps, you can create solid sweeps by moving a tool body along a path.

Tangent arc

An arc that is tangent to another entity, such as a line.

Tangent edge

The transition edge between rounded or filleted faces in hidden lines visible or hidden lines removed modes in drawings.

Task Pane

Located on the right-side of the SOLIDWORKS window, the Task Pane contains SOLIDWORKS Resources, the Design Library, and the File Explorer.

Template

A document (part, assembly, or drawing) that forms the basis of a new document. It can include user-defined parameters, annotations, predefined views, geometry, and so on.

Temporary axis

An axis created implicitly for every conical or cylindrical face in a model.

Thin feature
An extruded or revolved feature with constant wall thickness. Sheet metal parts are typically created from thin features.

TolAnalyst
A tolerance analysis application that determines the effects that dimensions and tolerances have on parts and assemblies.

Top-down design
An assembly modeling technique where you create parts in the context of an assembly by referencing the geometry of other components. Changes to the referenced components propagate to the parts that you create in context.

Triad
Three axes with arrows defining the X, Y, and Z directions. A reference triad appears in part and assembly documents to assist in orienting the viewing of models. Triads also assist when moving or rotating components in assemblies.

Under defined
A sketch is under defined when there are not enough dimensions and relations to prevent entities from moving or changing size.

Vertex
A point at which two or more lines or edges intersect. Vertices can be selected for sketching, dimensioning, and many other operations.

Viewports
Windows that display views of models. You can specify one, two, or four viewports. Viewports with orthogonal views can be linked, which links orientation and rotation.

Virtual sharp
A sketch point at the intersection of two entities after the intersection itself has been removed by a feature such as a fillet or chamfer. Dimensions and relations to the virtual sharp are retained even though the actual intersection no longer exists.

Weldment
A multibody part with structural members.

Weldment cut list
A table that tabulates the bodies in a weldment along with descriptions and lengths.

Wireframe
A view mode in which all edges of the part or assembly are displayed.

<u>Zebra stripes</u>

Simulate the reflection of long strips of light on a very shiny surface. They allow you to see small changes in a surface that may be hard to see with a standard display.

<u>Zoom</u>

To simulate movement toward or away from a part or an assembly.

Bug-Bot designed and rendered using SOLIDWORKS 2023

Index